PHILOSOPHY OF LOGIC

Other interview books from Automatic Press ♦ⱲP

Formal Philosophy
edited by Vincent F. Hendricks & John Symons November 2005

Masses of Formal Philosophy
edited by Vincent F. Hendricks & John Symons October 2006

Philosophy of Technology: 5 Questions
edited by Jan-Kyrre Berg Olsen & Evan Selinger February 2007

Game Theory: 5 Questions
edited by Vincent F. Hendricks & Pelle Guldborg Hansen April 2007

Philosophy of Mathematics: 5 Questions
edited by Vincent F. Hendricks & Hannes Leitgeb January 2008

Epistemology: 5 Questions
edited by Vincent F. Hendricks & Duncan Pritchard September 2008

Philosophy of Medicine: 5 Questions
edited by J. K. B. O. Friis, P. Rossel & M. S. Norup September 2011

Intellectual History: 5 Questions
edited by Morten Haugaard Jeppesen, Frederik Stjernfelt & Mikkel Thorup May 2013

The History of Logic in China: 5 Questions
edited by Fenrong Liu & Jeremy Seligman September 2015

Science and Religion: 5 Questions
edited by Gregg D. Caruso March 2014

Peirce: 5 Questions
edited by Francesco Bellucci, Ahti-Veikko Pietarinen & Frederik Stjernfelt July 2014

Social Epistemology: 5 Questions
edited by Duncan Pritchard and Vincent F. Hendricks, January 2015

Images: 5 Questions
edited by Aud Sissel Hoel, Peer Bundgaard and Frederik Stjernfelt, Febuary 2016

See all published and forthcoming books in the 5 Questions series at
www.vince-inc.com

PHILOSOPHY OF LOGIC: 5 QUESTIONS

EDITED BY

THOMAS ADAJIAN

TRACY LUPHER

Automatic Press ◆ ⊬P

Automatic Press ♦ $\frac{V}{I}$ P

Information on this title: www. vince-inc. com

© Automatic Press / VIP 2016

This publication is in copyright. Subject to statuary exception and to the provisions of relevant collective licensing agreements, no reproduction of any part may take place without the written permission of the publisher.

First published 2016

Printed in the United States of America
and the United Kingdom

ISBN-10 / 87-92130-56-9
ISBN-13 / 978-87-92130 56-3

The publisher has no responsibilities for the persistence or accuracy of URLs for external or third party Internet Web sites referred to in this publication and does not guarantee that any content on such Web sites is, or will remain, accurate or appropriate.

Cover design by Vincent F. Hendricks
Cover image: Jess Burns

Contents

Preface	vii
Acknowledgements	ix
1. Jc Beall	1
2. John Lane Bell	13
3. Johan van Benthem	21
4. Patricia A. Blanchette	25
5. Otávio Bueno	37
6. James Cargile	49
7. Mark Colyvan	57
8. Newton Carneiro Affonso da Costa	65
9. Pascal Engel	77
10. Susan Haack	87
11. Jaakko J. Hintikka	99
12. Dale Jacquette	105
13. Penelope Maddy	119
14. Lawrence S. Moss	123
15. Catarina Dutilh Novaes	129
16. Ahti-Veikko Pietarinen	139
17. Graham Priest	153
18. Stephen Read	161
19. Nicholas Rescher	171
20. Stewart Shapiro	177
21. Peter Simons	185
22. Timothy Williamson	195
23. Jan Woleński	207
About the Editors	219
Index	221

Preface

◆

Logic is perhaps the most fundamental tool for doing philosophy. It has made great advances since the nineteenth century. These advances raise new issues and reinvigorate enduring ones. What if anything are the defining features of logic? What is logical consequence? Is logic normative? What is the relationship between logic and the other sciences? Between logic and psychology? What is the appropriate methodology for investigating questions in the philosophy of logic? What continues to be learned from the founders of modern logic, Peirce and Frege? Is there 'one true logic'? If so, what is it? Or should we be 'pluralists' about logic? If so, what does that mean? How are the various paradoxes to be resolved? What ontology is required for logic? What is the status of the law of non-contradiction? New technical advances are often made in response to philosophical questions. The new developments raise new philosophical questions which in turn inspire further technical advances. And the cycle continues.

This book features interviews with leading logicians and philosophers of logic. Newcomers to the philosophy of logic may be surprised by the rich diversity of views. Each author was given the same 5 broad and suggestive questions.

1. Why were you initially drawn to the philosophy of logic?
2. What are your main contributions to the philosophy of logic?
3. What is the proper role of philosophy of logic in relation to other disciplines, and to other branches of philosophy?
4. What have been the most significant advances in the philosophy of logic?
5. What are the most important open problems in philosophy of logic, and what are the prospects for progress?

Some authors answered the five questions without comment; others criticized their presuppositions. Some authors followed the format closely; others did not. Some authors' contributions make significant technical demands on the reader, but many do not. Some contributions are more technical than others. The brief intellectual biographies and conversational tone will appeal to both students and scholars alike. We hope this book will stimulate students to dig more deeply into the philosophy of logic and, perhaps, contribute to the continuing discussion.

Acknowledgements

The editors are pleased to thank editor-in-chief Vincent F. Hendricks of Automatic Press ♦$\frac{V}{I}$P for coming up with the excellent idea of this series, and for his advice on this book. Associate editor Henrik Boensvang's feedback and help with the manuscript made the process painless. The editors thank Dean David Jeffrey and Alan Kirk for their financial support and encouragement of James Madison's Logic and Reasoning Institute. Chris Runyon's yeoman's work for the Logic and Reasoning Institute deserves special thanks. We would also like to thank the institute's student assistants Marie Eszenyi and John Gardner for their help.

September 2015
Thomas Adajian and Tracy Lupher,
Co-directors, James Madison University's
Logic and Reasoning Institute
Editors

1

Jc Beall

Board of Trustees Distinguished Professor of Philosophy at the University of Connecticut (Storrs CT USA), Professor of Philosophy at the University of Tasmania (Hobart TAS AUS) and Associate Research Fellow at the University of St Andrews (Fife Scotland UK).

---

1. Why were you initially drawn to the philosophy of logic?

I don't know why I was drawn, but I was first drawn to logic, and in turn philosophy logic.

For setting terminology, it may be useful to give some very sweeping and basic remarks on logic and its partner – the philosophy of logic. This can serve as background to the other questions.

Logic is a necessary truth-preservation relation over our language; it is the one that obtains only in virtue of 'logical vocabulary'. Exactly what counts as logical vocabulary is hard; and it is a driving question in the philosophy of logic. And exactly how we best answer the question is equally hard and also a central (though methodological) question in philosophy of logic.

An answer that I find to be useful – and which I endorse, as far as it goes – points to a very familiar tradition and topic-neutrality: the logical vocabulary is topic-neutral; and the traditional set of first-order vocabulary (at least without identity) is a good candidate for *logical* vocabulary. (All of this assumes a language. Throughout, I assume a common language, say, English – or at least some sufficiently simplified version of it.)

So-called logics of necessity, or 'logics of knowledge', or 'logics of obligation', or so on are one and all only so called. None of the given notions (e.g., necessity, knowledge, etc.) are topic-neutral in the required sense. This distinction is hard to precisely define (given the difficulty in defining *topic-neutrality*, etc.), but it is sufficiently clear and sufficiently familiar to be useful. Without the distinction, it begins to look as if every notion is a logical one, and that all theoretical pursuits – when done at a sufficiently rigorous or 'formal' or abstract or mathematical fashion – are in fact activities within logic. But that takes things too far.

A useful way to think of the logic versus non-logic divide is in terms of closure operators (familiar from Tarski and others). Think of *theories* in the common philosophical sense: a theory (in a language) is a set of sentences (from that language). What we want is to close our theories via 'absolute' or 'necessary' operators. And this is where a central role of logic (and other so-called 'logics of x') come into play. In particular, *logic* is the universal closure operator – the base operator, if you will – for all of our theories. All closure operators subsume the logical closure operator – subsume logic (qua weakest of the target 'absolute' subrelations). But logic is very weak in the sense that it concerns only the foundational, topic-neutral vocabulary. What we want, when we are constructing and expanding our theories via closure, are stronger closure relations that accurately reflect the (topic-relative) behavior of non-logical vocabulary. In fact, this is what we are doing when we do (as it's called) the 'logic of knowledge' or the 'logic of necessity'. We are in fact simply constructing the appropriate (the correct, adequate, etc.) closure operators for such notions.

Example: think of theory of knowledge, with K the (non-logical) target operator. Logic says nothing about K that it doesn't say about every notion whatsoever: it ignores K and speaks only of its behavior with respect to logical vocabulary (negation, disjunction, etc.). We need a stronger closure operator for any adequate theory of knowledge. Hence,

$$K\varphi \vdash_T \varphi$$

we construct our theory T's closure operator $\vdash_T$ by adding (non-logical) rules, such as the 'release' behavior of knowledge:
Such rules, conceived model-theoretically, have the effect of restricting the class of theory T's models. And it is the resulting class of models over which the target closure operator – the target 'necessary' or 'absolute' relation – is defined. This closure operator is not *logic*; but it subsumes and builds on logic, which, as above, is the universal, base closure operator for all of our theories.

On the foregoing way of thinking about logic, there is much room for logical theorizing. Logical theories are the theories that talk about the universal (necessary) truth-preserving behavior of logical vocabulary.[1] So-called classical logic reflects a theory according to which logical vocabulary behaves one way; various subclassical logics reflect a weaker

---

[1] Let me pause to highlight a huge issue in my (intended-to-be) broad and basic presentation: proof-theoretic versus truth-/model-theoretic accounts of consequence. I find it much easier to present things in the latter terms, but nothing that I say is intended to be in major conflict with a proof-theoretic account of logic – a proof-theoretic account of the target consequence relation or, better, universal closure operator.

account of such vocabulary. These differences come out vividly in the given closure operators. Example: the classical closure operator delivers all sentences into your theory if there's any negation inconsistency, while certain subclassical closure operators (viz., so-called paraconsistent operators) don't.

Exactly which account of the logical vocabulary (i.e., which logical theory) is the right account – or whether there can be more than one right account (given a language) – is the central question in the philosophy of logic with which I began these remarks. And it is a good place to stop, and turn to the next question.

2. What are your main contributions to the philosophy of logic?

My main contributions to the philosophy of logic involve the advancement and defense of non-classical logic in philosophy. Two examples of this work are my *Logical Pluralism* (OUP, 2004) with Greg Restall, and *Spandrels of Truth* (OUP, 2009), and another is my current project *Logic without detachment* (to appear with OUP), which advances and defends a strictly *subclassical* logic for truth theory. I will briefly say something about each work. (I also hope that some of my textbooks have been useful to philosophers interested in key ideas in philosophical logic; but I won't discuss these beyond mentioning them, namely, *Possibilities and Paradox: An Introduction to Modal and Many-Valued Logic* with Bas van Fraassen, and also *Logic: The Basics*.)

Logical Pluralism. This is the view that a single language can – and, in the case of our (say, English) language, does – enjoy a plurality of different consequence relations, that is, different logics. Some of the logics are paraconsistent (whereby arbitrary ψ fails to follow from arbitrary φ and $\neg\varphi$); some are paracomplete (in the sense that arbitrary ψ fails to imply either arbitrary φ or $\neg\varphi$); and some are neither. This elementary view continues to receive attention and criticism; and the prospects of an interesting and important logical pluralism remain under debate.

Spandrels of Truth. This is an account of how we enjoy a so-called transparent truth predicate, a transparent usage of 'true' whereby, for suitable names $\langle\varphi\rangle$ of sentences φ, $\langle\varphi\rangle$ *is true* and φ are everywhere-non-opaque intersubstitutable with each other. The account I give is a 'glutty' one, whereby familiar paradoxes (e.g., the liar) – the spandrels of 'true' – are true falsehoods, truths with true negations. In giving a transparency theory of truth, I join a tradition of truth theorists going back to early deflationists (Frank Ramsey, Paul Horwich, among others), though most explicitly (qua transparency) advocated by Hartry Field. In giving a glutty account, I join a camp of theorists going at least back to Florencio Asenjo, along with Chris Mortensen, Graham Priest, Richard Routley, and others – with Priest perhaps the most famous advocate of glut theory, at least in philosophy. But the account I give is very modest with respect to gluts: the only gluts are the 'spandrels of truth', the results of bringing in our transparent truth predicate. Were it not for our practical decision to introduce the truth predicate, there would be no gluts whatsoever; and the only gluts that do exist are ones in which 'true' is ineliminable. This account provides a simple way in which the standard view that eschews gluts – eschews negation-inconsistent theories – is mostly right; it's just the peculiar side effect of introducing 'true' that makes the general, no-gluts-at-all view incorrect.

LOGIC WITHOUT DETACHMENT.

The philosophical account of truth in *Spandrels of Truth* is correct (I think); but the overall account of the underlying logic was too complicated. The tricky problem for non-classical truth theories (or, for that matter, classical truth theories) comes with a detachable conditional – a conditional for which modus ponens is (logically) valid. (For the main issue, see discussions – easily accessible – of Curry's paradox.) A simple and natural paraconsistent logic may be achieved by weakening classical logic; the logics called 'FDE' (or 'first-degree entailments' or 'tautological entailments') and 'LP' (for 'logic of paradox' or, as in Asenjo, 'calculus of antinomies') are such logics. In my previous work, I endorsed LP as the basic first-order logic, but then – following a long-standing tradition – went on a quest to find a suitably detachable conditional. But this complicates the philosophical and logical picture more than it needs to be – or so I now think. My current project is to advance the idea that we have no detachable conditional – at least no conditional which is logically detachable (i.e., obeys modus ponens according to logic). This project faces an immediate challenge: how to explain (or explain away) our apparent use of modus ponens in rational theory construction. These ideas are the focus of a current book project, and some of the ideas are available in papers. (Examples: 'Free of detachment' in *Noûs*, 'Shrieking against gluts' in *Analysis*, and 'LP, K3, FDE and their classical collapse' in *Review of Symbolic Logic*.)

3. What is the proper role of philosophy of logic in relation to other disciplines, and to other branches of philosophy?

Logic is about what follows from what in virtue of logical vocabulary; and this relation is logical consequence or logical entailment or logical validity or, in a word, logic. The philosophy of logic, like any *philosophy of x*, raises standard philosophical questions about logic, about the target relation of logical consequence. Such questions are standard across philosophical subfields: what is the epistemology, ontology, metaphysics, normative status of the relation? (And other standard topics can and are raised.) In this sense, I do not think that the philosophy of logic has any special status in relation to other branches of philosophy, except that perhaps its target phenomenon (viz., logical consequence) is the weakest – broadest – constraint on theorizing in other subfields; it is the base or foundation of other (non-logical) theoretical closure operators on our theories.

On the other hand, there is significant interaction between philosophy of logic and other branches of philosophy. The philosophy of logic and other standard subfields have much in common, such as metaphysics (or at least 'formal' metaphysics) and philosophy of language. Example: one major topic in philosophy of logic concerns the appropriate level of logical analysis. In propositional logic, one only looks at the sentential level as the appropriate level of analysis; but logicians generally agree that logical validity demands a dive into the atomic innards – for example, names and predicates (and, of course, at least object variables if not predicate variables). It is here where discussions in (say) philosophy of language and philosophy of logic directly intersect. In particular, the behavior of names can make a big difference to logical validity, in particular the (in-) validity of various quantifier patterns. (This is why debates about 'free logic' in logical studies has been of direct interest to philosophers of language, and vice versa.)

Of course, sometimes, one is well-versed in logic and directly applies the formal picture to debates in metaphysics, philosophy of language, and philosophy of logic. Example: if one were well-versed in the standard (sometimes called 'Kripke') model theory of normal modal logics, one could simply take a face-value reading of the given model theory – the formal 'semantics' – and have interesting things to say in the metaphysics of possible worlds (e.g., Kripke, Lewis) or the philosophy of language (e.g., 'rigid designators' à la Kripke, Kaplan, and others). On a merely practical level, philosophers – perhaps especially graduate students in philosophy – would benefit from a rigorous study of standard model theories – formal 'semantics' – of standard logics, including both subclassical, anti-classical, and various model logics (e.g., epistemic, deontic, etc.). On a practical level, such study often opens up new philosophical views that are not easily seen except via a stark formal picture, such as the pictures delivered by standard model theories.

4. What have been the most significant advances in the philosophy of logic?

I will note some significant advances, staying neutral on whether they're the most significant advances – a question that would be hard to answer.

Applications of non-classical logic. One significant advance is the recognition and embrace of non-classical logics in philosophy. Great work continues to be done by a host of classical-logic driven philosophers of logic (e.g., Timothy Williamson, Brian Weatherson, Roy Sorensen, Stewart Shapiro, Kevin Scharp, Marcus Rossberg, Greg Restall, Rayo Agustin, Øystein Linnebo, Hannes Leitgeb, Volker Halbach, Michael Glanzberg, among many, many, many, many others); but there is a rising interest in philosophical applications of non-classical logic (e.g., myself, Roy Cook, Aaron Cotnoir, Catarina Dutilh Novaes, Elena Ficara, Hartry Field, Kit Fine, Leon Horsten, Ole Hjortland, Dom Hyde, Carrie Jenkins, Ed Mares, Julian Murzi, Graham Priest, Stephen Read, Greg Restall, Dave Ripley, Gemma Robles, Gill Russell, Lionel Shapiro, Zach Weber, Nicole Wyatt, Elia Zardini, and many others).[2] Many of the salient applications concern familiar paradoxes; however, there is much work that goes well beyond paradoxical phenomena into areas of metaphysics – for example, truth-making, grounding, conceptions of time (involving 'gaps' and 'gluts'), and more. While it's more of a sociological than conceptual shift, the full recognition, advancement, and acceptance of applying non-classical logics in philosophy is a significant advance in the field.

Rise of Substructuralism. One fairly recent advance concerns work on substructural logics, where (let me stipulate) this involves logics that give up one or more of the standard (classical) structural rules. Much of this activity builds on work from early so-called relevance logicians, though not all of the work reflects a commitment to (or achievement of) relevance logic. Instead, the idea is that our best overall logical theory is one according to which logic fails to obey some of the standardly assumed structural features (e.g., it might be thought not to be generally transitive, or might not 'contract' in standard fashions). I will not go into the details of this work here; but it is without question currently one of the main areas of research in logical studies and the philosophy of logic; and the work is clarifying a great number of buried assumptions that have been made by a great many philosophers of logic over the last few decades (or more). (For a look at some of the recent flurry of activity in this area – namely, philosophical applications of substructural logics – see recent work of, among many others, Paul Egre, Ole Hjortland, Ed Mares, Julian Murzi, Francesco Paoli, Stephen Read, Greg Restall, Dave Ripley, Lionel Shapiro, Heinrich Wansing, Elia Zardini, and others.)

Distinction between negation and rejection. While it is not new, the

---

[2] The parenthetically noted people are very far from either an exhaustive or representative list; but I note those whose current work is applying either classical logic or non-classical logic in new and increasingly influential directions. Omissions from the list does not in any way indicate that work is not notable or increasingly influential! (Restall, being a logical pluralist, counts as both classical and non-classical.)

distinction between negation qua logical connective and rejection qua mental state is now fairly broadly recognized. Non-classical theorists of negation have long stressed the distinction between rejecting a sentence and accepting its negation. In so-called paraconsistent logics, which deliver closure operators for (negation-) inconsistent but non-trivial theories (i.e., not every sentence is in the closure of the theory), you might accept both φ and $\neg \varphi$ in your theory, but you don't thereby reject φ. (It is generally thought to be metaphysically – or at least physically, mentally – impossible to both accept and reject the same sentence.) Dually, so-called paracomplete theorists of negation, who think that our (prime) theories can be closed under logic while nonetheless being negation-incomplete (i.e., some sentence and its negation fail to be in the logical closure of our true theories),[3] sometimes reject both $\neg \varphi$ and φ in their theories without thereby also accepting $\neg \varphi$. One important effect of recognizing the distinction between negation and rejection (and, dually, acceptance and the null operator) is that philosophers of logic increasingly recognize (or should recognize) the importance of a theory of reasoning (e.g., rational acceptance/rejection behaviors, patterns, etc.) as distinct from logic. This is related to another significant advance in philosophy of logic.

Distinction between reasoning and logic. Closely related to (though distinct from) the distinction between negation and rejection is the distinction between a theory of reasoning qua rational change in view (à la Gilbert Harman) and logic as a constraint on rational change in view. While the details of the distinction remain controversial, the basic distinction is generally acknowledged. And this is an advance in the philosophy of logic. For one thing, we can continue to think of logic as (let me say) truth-preservation over relevant possibilities in virtue of logical vocabulary, and so think of logic as perfectly 'absolute' and 'monotonic' and 'non-defeasible' and so on. On the other hand, we can (rightly) think of rational change in view (e.g., acceptance/rejection behavior) as very much 'non-monotonic', 'defeasible', and so on. Logic doesn't tell us what to accept or what to reject. Logic is used by a theory of rational change in view (a theory of reasoning) as a constraint on rational change in view: you can't rationally change your view to something that logic deems to be an invalid pattern. (Example: given the validity of $\varphi \wedge \psi \therefore \varphi \vee \psi$, our theory of reasoning tells us that it's irrational to exhibit an acceptance/rejection pattern whereby you accept $\varphi \wedge \psi$ while rejecting $\psi \vee \psi$.) The interplay between logic and its traditional role of constraint on rational change-in-view behavior is nicely illustrated by multiple-

---

[3] A prime theory is a theory – set of sentences – that contains a disjunction $\varphi \vee \psi$ only if it contains at least one of the disjuncts.

conclusion logic, but I leave this aside here. (Some of my recent work on living with non-detachable logics makes use – at least illustrative if not essential use – of multiple-conclusion subclassical logics. Other theorists, such as Dave Ripley and Greg Restall, go much further in this direction: they *define* logic in terms of acceptance/rejection behavior – a radical though intriguing program, by my lights.) Exactly how important the distinction may be depends on the philosophical program in question; but it is an advance in the philosophy of logic that the distinction is now clearly acknowledged.

Distinction between logical truth and logical consequence. Another advance concerns another old distinction which is now fairly broadly appreciated: namely, the distinction between logical consequence and logical truth (or, if you prefer a proof-theoretic version, a valid deduction and a theorem). In classical (and many other) logic(s), we have a simple deduction theorem that aligns logical consequence (here, the turnstile) and the logical truth of a certain sort of sentence (here represented via an arrow):

$$\varphi_0, \varphi_1, \ldots, \varphi_n \vdash \psi \text{ iff } \vdash \varphi_0 \wedge \varphi_1 \wedge \ldots \wedge \varphi_n \to \psi.$$

When one has this sort of deduction-theorem link between consequence and the logical truth of some sort of sentence, one can often slip into thinking of logic (qua discipline) as concerned only with logical truths (or, from a proof-theoretic angle, theorems). But it's an advance in the philosophy of logic that the distinction between consequence and logical truth is now firmly recognized by philosophers of logic. The distinction is very clear in non-classical logics, including very simple subclassical logics. Example: in Strong Kleene K3, the 'link' between logical truth of a material conditional and consequence fails, since φ implies itself in K3 but $\varphi \to \varphi$ is not logically true, at least where the arrow is the material conditional defined per usual as the disjunction of the consequent and negated antecedent. Similarly, in fact dually, in the case of the paraconsistent subclassical logic often called 'LP' – for 'logic of paradox' or, as in Asenjo's presentation, 'calculus of antinomies' – we have the logical truth of $(\varphi \wedge \neg \varphi) \to \psi$ but not the corresponding implication (i.e., not the corresponding valid argument), at least where, again, the arrow is the defined material conditional.

This distinction – between consequence and logical truth – is now widely recognized by philosophers of logic, though the philosophical issues arising from the distinction remain very much under debate.

Logic as another theory. One more significant advance, at least since Quine, is that one's logical theory – one's theory of logical consequence – is as much a theory as any other theory, subject to the same epistemological and metaphysical problems as other theories. One difference between logical theories and other theories has been noted throughout my remarks: namely, that logical theory deals with the weakest, broadest closure operator for all of our theories, unlike the relation(s) involved in other (non-logical) theories. But it remains an advance that philosophers of logic now accept that logical theory is theory – our rafts are afloat even more than we might hope. (I note that Quine himself seemed to be slightly confused about the reach of this otherwise very Quinean lesson. In particular, as is familiar, Quine seemed to dismiss a variety of non-classical logics as 'changing the subject', even though the main Quinean lesson – the Quine-the-good lesson, so to speak – is that the subject matter is the validity relation, and our theories of that relation may range from 'classical' to subclassical to anti-classical.)

5. What are the most important open problems in philosophy of logic, and what are the prospects for progress?
I list important open problems, but remain neutral on whether they're the most important such problems – a difficult matter to assess.

Problem: Logic versus other entailment relations. An old but still-open problem is the difference between logical entailment – logical consequence, logical validity – and other entailment relations, other necessarily truth-preserving relations. I have relied on a traditional answer: logical entailment is entailment in virtue of logical vocabulary. I believe that that's right, but it leaves open the criterion of logical vocabulary – an open problem. As above, I myself invoke tradition and a simple (possibly too simple) notion of topic-neutrality: the logical vocabulary is those expressions to which tradition has generally pointed as topic-neutral logical vocabulary. One way of understanding (though not precisely defining) this claim is, as above, in terms of closure operators on our theories: logical vocabulary is the vocabulary involved in our weakest, universal closure operator – the operator at the 'bottom' of all of our other theoretical closure operators. The standard first-order vocabulary is a good candidate for logical vocabulary so understood, but the issue is difficult. It would be a major accomplishment if we could find an uncontroversial criterion for logical vocabulary. The prospects on this score are not clear; but I see no reason to be pessimistic. For now, theorists need to specify what they take to be logical vocabulary, spell out their theories (and target closure operators built on top of the specified logical closure operator), and then let the theories be measured comparatively in terms of standard theoretical virtues.

Problem: Status of truth as logical. This question is directly related to the previous open problem, but I think it worth flagging independently. That there is a simple, so-called transparent usage of 'true' (whereby, for suitable names $\langle\varphi\rangle$ of sentences φ, $\langle\varphi\rangle$ *is true* and φ are intersubstitutable in all non-intensional and non-opaque contexts) is a commonly recognized view, if not yet widely embraced. Such a usage is one in which truth is 'deflationary', indeed perfectly 'see-through' given said transparent or intersubstitutable behavior. But a question immediately arises: is the truth predicate, so understood, a *logical* expression? It would seem to be perfectly 'topic-neutral' in any standard sense. Philosophers of logic often describe a transparent truth predicate as – at least analogously – 'logical', in the way that the standard (first-order) quantifiers are thought to be logical, or conjunction, or etc. But is truth, so understood, really logical? Does anything important (or interesting) hang on this? This is an open problem the importance of which is itself an open problem. I think that the prospects for resolving it are fairly bright: it simply needs to be carefully discussed by the community of philosophers of logic. (NB: precisely the same question emerges for other relations, such as predication, denotation, etc.)

Problem: Relation between exemplification/properties and membership/sets. There is a difference between exemplification and membership. The former motivates a principle of unrestricted comprehension: for any meaningful predicate $\varphi(x)$, there is a (say, property) exemplified by all and only the objects y of which $\varphi(x)$ is true. Membership does not motivate unrestricted comprehension, at least if standard (say, ZF/C) set theory is to be a guide to the membership relation. And there is another difference: properties can be distinct while having the same extension – even, perhaps, necessarily the same extension – and, so, do not motivate an identity-by-extension criterion; but sets, at least if standard set theory is a guide, are essentially identified by extension (so-called extensionality principle).

The distinction between exemplification and membership is not at all new. (According to Myhill, Kurt Gödel invoked the distinction to claim that set theory never faced paradox; the paradoxes, such as Cantor/Russell/Zermelo, plagued *property* theory, not set theory.) But what remains an open problem in the philosophy of logic is the relation between properties and sets, the relation between exemplification and membership. Finding a plausible account of exemplification, membership, and their relation is an important and wide-reaching problem: it has consequences in both philosophy of language and metaphysics, if not more widely. The prospects for success on this problem are good, I think; it's a problem that has lurked at the edges of mainstream philosophy of logic, but has not – perhaps until recently – gained firm attention from a wide swath of the community.

Problem: Philosophical understanding of substructuralism. As noted above, there is a recent flurry of activity around substructuralist accounts of logic and standard paradoxes. This is an exciting advance in the field of philosophy of logic; however, coming to grips with an intuitive account of substructuralism is a pressing open problem. What is needed is a plausible philosophical picture that clearly motivates the failure of standard structural rules such as contraction or cut (or more!). The prospects for achieving such a philosophical picture are unclear, but there is no reason to be pessimistic. The focused application of substructuralism to problems of familiar paradoxes is very recent – or at least only very recently widely recognized and pursued.

Problem: Philosophical significance of dual theories. While there mightn't be a well-defined general notion of duality, the notion is clear enough when it comes to various logics. There is an under-explored question in the philosophy of logic: are there any advantages that a glut theory enjoys over a dual gap theory, or any problems that one faces that the other doesn't. (This problem can be more precisely defined by looking at standard dual logics – like the subclassical K3 and LP logics – and their chief philosophical applications, especially transparent truth theories along Kripke, Martin-Woodruff, Dowden, et. al. lines.) In the one case, a theory contains both the 'glutty claim' $\varphi \wedge \neg \varphi$ and the 'gappy claim' $\neg \varphi \wedge \neg \neg \varphi$, while the other (say, the K3-based theory) contains neither claim. It would seem that any theoretical virtue that one theorist enjoys the other has a (dual) virtue, and similarly for any alleged problems. But the issue has not enjoyed sustained exploration: can there be philosophical advantages of a theory over its exact dual? It would be useful to have a sustained discussion of this problem by the wider community in philosophy of logic. I suspect that the prospects for discovering something interesting and important are bright.

Problem: deduction-theorem links. One other notable open problem concerns logics in which there is no deduction-theorem link between consequence (the logic) and the logical truth of a given sentence (or, to focus, a given connective, often a conditional-like connective). Need there be such a link in our best account of logic? If so, what form need the link take? These questions have been discussed in the past, in the context of early work in relevance logic (e.g., Anderson, Belnap, Brady, Dunn, Steve Read, and others); but the increasing philosophical applications of non-classical logics call for an earnest return to the question. I remain optimistic that the community of philosophers of logic can make clear progress on the question. I hope so. (For background on why this 'link' can fail, see again standard discussions of Curry's paradox and, e.g., truth theories.)

Bibliography

Jc Beall. *Spandrels of Truth*. Oxford University Press, Oxford, 2009.

Jc Beall. "Free of detachment: logic, rationality, and gluts". *Nous*, 2014, forthcoming.

Jc Beall. "Multiple-conclusion LP and default classicality". *Review of Symbolic Logic*, 4(2):326–336, 2011.

Jc Beall. "LP+, K3+, FDE+ and their classical collapse". *Review of Symbolic Logic*, 2012.

Jc Beall. "Shrieking against gluts". *Analysis*, 2013, forthcoming.

Jc Beall. *Logic: The Basics*. Oxford: Routledge, 2010.

Jc Beall and Greg Restall. *Logical Pluralism*. Oxford: Oxford University Press, 2006.

Jc Beall and Bas van Fraassen. *Possibilities and Paradox: An Introduction to Modal and Many-Valued Logic*. Oxford: Oxford University Press, 2003.

2

John Lane Bell

Professor of Philosophy
University of Western Ontario

---

1. Why were you initially drawn to the philosophy of logic?
I should say at once that I have worked chiefly in mathematical logic and the foundations and philosophy of mathematics, and my interest in the philosophy of logic (but not in philosophy *per se*) is the result of my activity in those areas. My route thereto was somewhat circuitous. In youth I was attracted to physics, especially relativity theory and cosmology—I actually attended one of Fred Hoyle's lecture courses on the subject in Cambridge in the early 1960s. (Parenthetically, I may mention that it was through Hoyle's lectures that I first heard the name Gödel, not of course in connection with his discoveries in logic, of which I was then wholly ignorant, but as the deviser of cosmological models containing closed timelike lines.) While I had become quite adept in mathematical physics, at some point it dawned on me that I had no genuine understanding of what I was actually doing. Accordingly I decided to turn away from physics, my first love, and concentrate on pure mathematics. While mathematics lacked, in my eyes, the romantic appeal of cosmology, it had the compensating merit that its concepts and methods could, in principle at least, be fully presented to the understanding. My flight to mathematics was fuelled by my discovery of John Kelley's classic work *General Topology*. Kelley also furnished my first introduction to set theory. This in turn led me to study Gödel's monograph, *The Consistency of the Axiom of Choice and the Generalized Continuum Hypothesis*. The first two-thirds of this mathematical tour-de-force, in which Gödel presents his axiom system for set theory and develops its essential properties, seemed reasonably clear. But, despite my best efforts, I was unable to fathom the final part of the work, its grand finale, so to speak, in which, accompanied by an inaudible clash of cymbals, the consistency of the generalized continuum hypothesis is established. A good few years were to pass before I felt I truly understood what was going on here.

Another influence was Bourbaki's Éléments *de Mathématique*. On first coming across some volumes of this monumental work in Blackwell's bookshop, I was excited to find that it was intended to be a complete, systematic account of abstract mathematics, precisely the kind of mathematics to which I had already been converted by Kelley's *General Topology*. The *oeuvre Bourbachique* included not only *Topologie Génerale*, but *Algèbre*, *Thèorie des Ensembles*, *Espaces Vectoriels Topologiques*, *Algèbre Commutatif*— magical titles in my eyes. I bought as many volumes as I could afford, often in obsolete—and so cheaper—editions (the whole enterprise seemed to be undergoing constant revision), and commenced to work my way through the collections of challenging exercises at the end of each section. I toiled mightily, in particular, to formulate solutions to the exercises on ordered sets in Chapter 3 of the *Thèorie des Ensembles*. It was from these that I first learned about ordinals, which Bourbaki presents in the original Cantorian manner as order types of well-ordered sets.

Kelley, Bourbaki, Gödel: it was through their influence that I was led to the foundations of mathematics. My interest in philosophy, on the other hand, derived from my being a voracious and eclectic reader. As an undergraduate I recall reading Plato's *The Last Days of Socrates*, William James's "Essays on Pragmatism", G. E. Moore's philosophical essays, Hegel's *Philosophy of History*, Descartes' *Discourse on Method*, Spinoza's *Ethics* (the statements of the theorems at least, since I found the "proofs" unenlightening), Leibniz's delphic *Monadology*, some Locke, Berkeley and Hume, Schopenhauer's *Essays in Pessimism*. And of course Bertrand Russell's breezily brilliant, if irresponsible, *History of Western Philosophy*. My attempts to penetrate the profundities of Kant's *Critique of Pure Reason* were frustrated by the work's apparent indigestibility. (I was only to appreciate its depth and philosophical importance many years later.) I greatly enjoyed Hans Reichenbach's *Philosophy of Space and Time*. On Blackwell's shelves in Oxford I came across Norman Malcolm's *Wittgenstein: A Memoir*. I was deeply moved by Malcolm's portrayal of Wittgenstein, in which he emerges as an intellectual ascetic of compelling moral grandeur. Wittgenstein's tiniest defiances of convention, for example, his refusal to wear a tie at dinner in Trinity College, I found admirable. Reading Malcolm's memoir stimulated me to attempt to read Wittgenstein's philosophical works. I was intrigued by the *Tractatus Logico-Philosophicus*, a masterpiece of sybilline refinement and compression in which Wittgenstein embarks on the heroic effort of reducing philosophy to the expressible, but in the end washes up on the shores of the ineffable. The conventionalism of the later Wittgenstein's *Philosophical Investigations*, I found less appealing.

I was drawn quite early on to the foundations of mathematics, and to general philosophy, but my conscious interest in the philosophy of mathematics per se was comparatively slow to crystallize. As an undergraduate I had developed an interest in set theory and the philosophy of the infinite in general. Later I came to believe that mathematics, while being, like art, a beautiful sublimation of human activity, has, in the last analysis, to be understood as the product of actual human beings living in the world. This led me to see that mathematics actually has a *hidden content*, which can actually be *argued about*. This is the opposite of the unthinking Platonism/realism to which I was, I guess, initially attracted as offering the simplest account of mathematical truth, and which also possessed the additional advantage of avoiding what I then felt to be a certain cynicism inherent in Formalism. (Still, as I have come to learn, Formalism has the great merit of offering the weary ex-Platonist a refuge.) But, like the child's loss of belief in Santa Claus, I came to regard the Platonistic account of mathematical entities as a kind of fairy tale, and in any case as engendering insuperable epistemological difficulties. I may parenthetically remark that I have since come to liken Platonism to a (necessary) disease, which, like measles, must have been contracted in one's youth so as to confer an immunity in later life.

A key stage in the development of my interest in the philosophy of mathematics came through my efforts to understand *topos theory*. I was very struck by Bill Lawvere's insight that a topos is an objective presentation of the idea of *variability*, and that its internal—intuitionistic—logic may be considered as a logic of variation. Later I went so far as to attempt to use the topos concept as the basis for a "local" (as opposed to "absolute") interpretation of mathematical statements. I suggested that the unique absolute universe of sets central to the orthodox set-theoretic account of the foundations of mathematics should be replaced by a plurality of local mathematical frameworks—elementary toposes—defined in category-theoretic terms. Observing that such frameworks possess sufficiently rich internal structure to enable mathematical concepts and assertions to be interpreted within them, I maintained that they can serve as local surrogates for the usual "absolute" universe of sets. On this account mathematical concepts will in general no longer possess absolute meaning, nor mathematical assertions (e.g. the continuum hypothesis) absolute truth values, but will instead possess such meanings or truth values only locally, i.e., relative to local frameworks. The absolute truth of set-theoretical assertions would then, I held, give way to the subtler concept of invariance, that is, validity in all local frameworks. Thus, e.g., while the theorems of constructive arithmetic turn out to possess the property of invariance, the axiom of choice or the continuum hypothesis do not, because they hold true in some local frameworks but not others.

I still find this view attractive, but it is, after all, only one among many possible accounts of mathematics. If I were pressed to characterize my present attitude towards the foundations of mathematics, I would use the word *pluralistic*: no unique foundation, rather an interlocking ensemble of "foundations". My pluralist attitude also extends to logic: instead of a single overarching Logic governing all forms of reasoning, my own experience has led me to conclude that each type of reasoning - classical, intuitionistic, quantum, linear - carries its own logic in the form of rules laid down in accordance with the nature of the objects or concepts being reasoned about.

2. What are your main contributions to the philosophy of logic?

While most of my work has been in technical mathematical logic, I have made some contributions to the philosophy of logic.

The first of these was essentially a contribution to philosophy of science. As an ex-aspiring-physicist I had long been intrigued by quantum theory, with its mysterious superpositions of states and incompatible measurements; and as a logician my curiosity was piqued by the so-called quantum logic, whose characteristic feature is that its algebra of propositions is not a Boolean or Heyting algebra, but a certain kind of nondistributive lattice—an ortholattice. All of these facts can be, and are, formally derived from the standard Hilbert space formalism of quantum theory. I became interested in the problem of formulating some simple principles, free of the technicalities of the theory of Hilbert spaces, from which one could derive the anomalous features of quantum theory, as well as the ortholattices underlying quantum logic. I came up with two approaches. The first, essentially topological, was based on the idea of using what I called a proximity space, a set equipped with a symmetric reflexive relation "close to". The lattice of parts of such a space is an ortholattice. There is a natural way, which I called "manifestation", similar to Paul Cohen's celebrated concept of set-theoretic forcing, of relating propositions (actually attributes) to parts of the space. The propositions manifested over the whole of every proximity space are (essentially) the theses of quantum logic. Given two propositions P, Q, their superposition can be identified with $\neg\neg(P \vee Q)$, and they are incompatible if there is a proximity space with a part manifesting P but not $Q \vee \neg Q$, or vice-versa.

In my other approach to the problem, I showed how to construct the ortholattices arising in quantum logic from what I saw as the phenomenologically plausible idea of a collection of ensembles subject to passing or failing various "tests". A collection of ensembles forms a certain kind of preorderd set with an additional relation I called an orthospace: I showed that the complete ortholattices, in particular those of quantum theory, arise as canonical completions of orthospaces in much the same way as arbitrary complete lattices arise as canonical completions of partially ordered sets. I also showed that the canonical completion of an orthospace of ensembles may be identified with the lattice of properties of the ensembles, thereby showing exactly why ortholattices arise in the analysis of "tests" or experimental propositions. I went on to axiomatize the concept of "test" itself in terms of the more primitive notion of "filters" acting on ensembles. "Passing" an ensemble through a filter s produces the subensemble of entities that have "passed" the test corresponding to s. Two filters s and t can be juxtaposed to produce the compound filter st, but in general $st \neq ts$. When the latter is that case, the two tests corresponding to s and t are, like position and momentum measurements in quantum theory, not simultaneously performable, that is, *incompatible*. When (and only when) $st = ts$, the juxtaposition of s and t corresponds to their logical conjunction. In this setting, it is the noncommutativity or incompatibility of filters or "tests" that gives rise to "quantum logic".

A philosophical problem which had long intrigued me was: why is traditional logic *bivalent*, that is, why is it assumed that there just two truth values rather than some other number? What is it about the number 2 that gives it this special position in logic? Wittgenstein seems to take the fact for granted when (in his *Notes on Logic*) he says that propositions have two "poles". It is often claimed that bivalent logic is the "logic of realism", that is, logic in which propositions are construed as referring to independently existing objects, in contrast with "anti-realist" logics such as intuitionistic logic (I don't agree that intuitionistic logic has to be thought of as anti-realist—but let that pass). However, this begs the question, since the thought immediately arises: what is it about the realm of independently existing objects that confers bivalence on propositions referred to it? Why shouldn't the number of objective truth values be, say, 3, like the number of spatial dimensions? Wittgenstein recognized the possibility of this question arising but simply dismissed it.

One way that occurred to me of explaining the role of the number 2 in logic is by moving from individual propositions to sets of propositions, or *theories*. Frege had suggested that the bivalence of the logic of concepts arises from their having *sharp boundaries*: one can determine with exactitude, for such a concept, when an object falls under it , or when it does not. In other words, a concept's possession of a sharp boundary means that the theory of the concept is complete with in regard to atomic propositions. It is then natural to extend this prescription to arbitrary propositions. So, metaphorically, we may say that (the concept determined by) a theory has sharp boundaries if it is *complete*, that is, if any proposition in the theory's vocabulary is provable or refutable from the theory. But it is well known that, for any complete theory T (in propositional intuitionistic or classical logic), it is possible to assign the *two* truth values 0, 1 to propositions in such a way as to respect the logical operations, and also to assign precisely the propositions in T the value 1. And conversely, if such a bivalent assignment exists, the theory is complete. That is, the number 2 is simply the numerical representative of completeness, or the possession of "sharp boundaries".

The major logical consequence of bivalence (although not equivalent to it) is the *law of excluded middle:* the assertion, for any proposition P, of the disjunction $P \vee \neg P$. This is of course the logical principle which whose affirmation distinguishes classical from intuitionistic logic. Like bivalence the law of excluded middle has been taken to be characteristic of logic in which propositions are construed as referring to independently existing objects. I found that, if one starts with intuitionistic predicate logic, and extends it to include Hilbert's ε-terms (these are essentially objects named by the use of the indefinite article: *a* such-and-such), then the law of excluded middle becomes provable. That is, the law of excluded middle is, after all, derivable from what can reasonably be construed as an ontological principle.

3. What is the proper role of philosophy of logic in relation to other disciplines, and to other branches of philosophy?

I think that the philosophy of logic should, and in fact does, play a dialectical role in relation to its sister disciplines, guiding them and, reciprocally, responding to their internal development. Let me attempt to illustrate what I mean. Cantor's philosophy of the infinite (and his associated, if lesser-known, championship of the reduction of the continuous to the discrete) played a major part in his development of set theory, which, as is well-known, came to permeate mathematics. Partly in reaction to the unrestricted use of Cantorian set theory in mathematics, Brouwer formulated his philosophy of intuitionism with its radical revision of the laws of logic (rejection of the law of excluded middle, etc.). This in turn led to a prolonged discussion of the nature of logic, culminating in Carnap's principle of tolerance, which has become a central doctrine in the philosophy of logic.

4. What have been the most significant advances in the philosophy of logic? Here, in my view, are some of those advances:

- Boole's analysis of logic in mathematical terms.
- Frege's analysis of propositions and the idea that truth values are the references of propositions.

- Russell's type-theoretic account of propositions.

- The emergence of intuitionistic logic and its challenge to the supremacy of classical logic.

- The development of logical pluralism, the idea that there is more than one Logic.

- The recognition, following Gödel's pioneering work, that many logical systems are incomplete.

- Tarski's analysis of the concept of truth - in particular, his undefinability theorem.

- The emergence of many-valued logic, the idea that propositions can possess more than two truth values.

- The idea, arising from topos theory, that logical connectives and quantifiers can be construed as mathematical operations on a domain of truth values.

- The "propositions-as-types" doctrine underlying constructive type theories such as that of Martin-Löf.

5. What are the most important open problems in the philosophy of logic, and what are the prospects for progress?

Here are some questions in the philosophy of logic that strike me as being of some importance (I find them interesting, at least).

Should the foundations of logic be monistic or pluralistic? In particular should quantum logic and other "deviant" systems be considered logic?

What is the meaning of a "true" contradiction? Can there be "objective" contradictions, as Hegel claimed, or "contradictions in nature", as Engels claimed?

Is it in the nature of judgments to be bivalent (true or false)? In particular, should bivalence ultimately prevail in metalogic?

What is the nature of propositions about the quantum microworld? In what respect do these propositions differ from those concerning "ordinary" objects? In particular, can microobjects such as photons be said to possess properties in the same sense as do "ordinary" objects?

It seems to me that some progress has already been made on all of these questions, and I confidently expect further advances to continue to be made.

3

Johan van Benthem

Professor of Logic
Stanford University, USA, and Tsinghua University, China

---

1. Why were you initially drawn to the philosophy of logic?

I am not a philosopher of logic: I must have missed the draw. Still, I have taught the subject, I know many practitioners, and as a working logician, I do have thoughts on it. And as we all know, being asked for one's views in print on anything is a heady drink to refuse. But what is the philosophy of logic? In my student days, a few American textbooks by Putnam, Quine, and Haack set the agenda, and, nowadays, there are also influential Australian and European schools. But despite the great names involved, the subject of the field is not crystal-clear, or its status beyond dispute.

2. What are your main contributions to the philosophy of logic?

I cannot claim any contributions to the philosophy of logic as normally understood. I do not doubt for a moment that logic can and should profit from philosophical reflection. Over the years, that is what I have tried to do on many occasions. But the key issues on my agenda bear little resemblance to what is common in the philosophy of logic. For a start, I find it very hard to relate to what seems to be the central question in the area: 'What is Logic?' Maybe this is due to my education. It was already my famous Amsterdam predecessor Evert Beth who pointed out, in his influential book, *The Foundations of Mathematics* (1959), that "what is" questions are typical for a traditional philosophy driven by what he saw as an outdated Aristotelian Theory of Science looking for the 'essence' of scientific fields. In opposition to this, Beth placed the modern understanding of sciences as developing entities with logical foundations, to be undertaken only by philosophers who have an intimate knowledge of the technical state of the art.

The starting point for my own thinking about major philosophical issues concerning logic has been the actual developments in the field. You can find some of the resulting themes in my article 'Wider Still and Wider, Resetting the Bounds of Logic' (*Topoi* 2001). They included

(a) questioning the standard methodology of formal systems, with a view to avoiding *system imprisonment* of the general insights that a discipline should provide, (b) a critical understanding of the notions of *invariance* underpinning logical languages, going back to the tradition of Helmholtz and Klein in the 19th century, and (c) rethinking the delicate balance of *contents versus wrappings* in the complexity of standard logical systems such as first-order logic whose sometimes accidental design details are taken as gospel by many philosophers nowadays. In recent years, I have become concerned with further foundational topics about logic at the same level of generality. One is the *product/process duality* that underlies so much of logic, and the resulting changes in our understanding of the spectrum of informational activities that a more dynamic logic can study. Another is charting the diversity of *notions of information* that occur in modern logic, a sort of microcosm of the diversity of notions of information generally, and how to best bring these in line. And a more recent interest is the widespread *implicit/explicit* contrast in designing logics for the same phenomenon, either by reinterpreting standard languages and inferences, obtaining deviant logics and new notions of consequence, or by adding new vocabulary conservatively to a classical base. With these general themes, I hope to have demonstrated that logic, like any serious scientific field, keeps generating new philosophical challenges as its notions and results keep growing.

What this view implies is that logic has dynamic frontiers, not static ones. Logic *is* what it is *becoming*, the same way we think about the development of physics, or to mention a more modern discipline, computer science. With respect to such disciplines, serious philosophers absorb the best current work, and raise exciting related issues. And the same should be true for logic. And this holds for historians as well as systematic philosophers. I find much of the recent interest in Bolzano (to which I have contributed a bit myself) and the Middle Ages (especially, studies of argumentation and debate) highly suggestive for taking a fresh look in our modern understanding of the field.

3. What is the proper role of philosophy of logic in relation to other disciplines, and to other branches of philosophy?

The most obvious living contact for the philosophy of logic should be with logic itself. But to me, themes in the philosophy of logic are often far out of touch with the field. For instance, it is customary to read that logic is 'essentially' about valid consequence: and everyone seems to think this is just an innocent commonplace. But this description of logic was already wildly out of touch around 1950, when logic had started branching out into proof theory, model theory, and theory of computation, each of them major logical topics in themselves. (Around 1960, Beth wrote a famous paper on 'constants of logical thought' where he discussed three major themes throughout the history of logic: proof, definition, and algorithm.) Another tendency that baffles me is the popularity of finding the 'boundaries of logic', an enterprise that philosophers do not attempt with any other scientific discipline that they study. Philosophers of logic keep looking for semantic notions of invariance, or inference formats that are supposed to delimit *exactly* what real logic is about, striving to create some Andorra of pure core territory, a tiny condominium of philosophy and mathematics kept safe from outside influences by the surrounding Pyrenees. Why this urge to fix boundaries? Why is it taken for granted that this is a feasible, let alone, an interesting enterprise? Of course, boundary setting does appeal to temperaments who enjoy sticking labels (especially when they can accuse someone of not 'doing real logic'). But to me, this seems an empty pastime.

Admittedly, I have never made a dent with the above views. A prominent philosopher of logic once said in a public discussion that, if what they studied did not reflect the current state of the art, then so much the worse for contemporary logic. It was the job of philosophers to define what logic is, and like Bradshaw's guide for Victorian travellers, tell the logicians what to do and where to stay. I found this response very revealing. It sounds like the old German Naturphilosophie that claimed a privileged philosophical access to what nature was really about, disjoint from the evil fumes of physics laboratories and hocus pocus of mathematical proofs. Of course, Naturphilosophie is dead. A related prescriptive tendency is the popularity of historicizing views of logic as culminating in the (largely mythical) topical and methodological purity of Frege's age.

Maybe all this is just ill-tempered and mean-spirited criticism. But I do think some of these habits can be harmful. There have been several a priori philosophical arguments claiming to show the impossibility of particular avenues of research, that have held up things for too long. Examples are Tarski's arguments about the inconsistency of natural language, or Harman's claimed incoherence of a logic of dynamic actions, and one could cite many others. And even when there is an attempt at getting closer to the realities of logic that I really like, such as the program of 'logical pluralism', another surprise awaits. The examples of legitimate broader consequence relations turn out to come from intuitionism (1931) and relevant logic (1960s). Now consider non-monotonic logic, easily the most exciting innovation in logic of the 1980s, with roots going back to Bolzano's and Peirce's pioneering work, whose repercussions are still felt through wide swaths of the academic world. As I learnt recently to my consternation, non-monotonic consequence is excluded from the pluralist program given what it counts as a legitimate consequence relation.

4. What have been the most significant advances in the philosophy of logic? and 5. What are the most important open problems in philosophy of logic, and what are the prospects for progress?

I see myself as making a plea for what some people nowadays call a *practice-based philosophy of logic*, close to the realities of the field, just as we would expect from the philosophy of any serious intellectual field. Still, I note that some 'practice-based' accounts fix the status quo of foundational research in the 1930s, making logic consist in mathematical theorem proving only. That makes it hard to understand the conceptual achievements of such non-theorem-proving leading logicians as Peirce, Frege, Bolzano, Carnap, Quine, or Hintikka. What I mean, by contrast, is an informed awareness of logic in its full breadth as it is in our own day and age. In particular, this means taking careful note of the conceptual advances made in philosophical logic, and in the logical foundations of computation, the largest segment of creative research in the field today.

To me right now, however, the most burning current issue is the interplay of *normative* and *descriptive* in logic. As I have explained in my 2008 paper 'Reasoning and Psychology: do the facts matter?', traditional 'barrier theses' like Russell's Misleading Form Thesis for natural language, or Frege's 'anti-psychologism' separating logic from the exciting cognitive psychology of his day, are beginning to dissolve. They are a sign of intellectual poverty rather than nobility to people who are really interested in studying reasoning, located as it is at the interface of what is eternally valid and what is done by us humans. Of course, there are arguments for these theses that once seemed convincing, but at pivotal periods in intellectual history, these are often just swept away by winds of change.

These issues may quickly become technical again, as they were in the golden age of foundational research that some philosophers of logic seem to have a longing for. In particular, I see major growing contacts and confrontations between two different styles of understanding information and reasoning (even as paradigms inside philosophy): *logic* and *probability*. This is not just a simple matter of mathematical interfacing. Typically, we need philosophers to think about what it would mean to combine these worlds in non-trivial harmonious ways. Indeed, there are logicians doing just this today, and in that sense, the true philosophers of logic in my sense may already walk amongst us under other labels.

4

Patricia A. Blanchette

Professor
Department of Philosophy, University of Notre Dame

---

1. Why were you initially drawn to the philosophy of logic?

In retrospect, I realize that I was drawn to the philosophy of logic long before I knew that there was such a field. When I was an undergraduate, before I had read any philosophy, I took a set theory course, and found myself intrigued by questions not covered in the class – questions like "why *these* axioms rather than some plausible alternatives?" and "what exactly are we trying to model, or to do, with a theory of sets?" I later took some courses in logic, and was again struck by questions not dealt with in the syllabus, questions like "Is there some reason to prefer a logic that satisfies the Löwenheim-Skolem theorems to one that doesn't?", "Is logic just a matter of 'relations of ideas'?" (I'd read some Hume by then), and "What exactly does it mean to say that number theory is 'part of' or 'reducible to' logic, and what does that claim imply with respect to the nature of mathematics?"

Though these questions (and others like them) intrigued me from early on, I was initially inclined to think that they were insufficiently precise to admit of clear answers, or even of clear proposals for answers, and so it didn't initially occur to me to try to do serious work on them. In graduate school, I thought at first that I'd work on the philosophy of language, and leave questions like the above for moments of idle musing.

Two things changed my attitude towards the reasonableness of pursuing serious work in the philosophy of logic. One of these was that I read John Etchemendy's *The Concept of Logical Consequence*, in which I found a beautiful example of an attempt to take seriously the question of how a particular, central pretheoretic notion (that of logical consequence) matches up with, or fails to match up with, a particular kind of modern account of that notion.[1] Etchemendy's generally-negative conclusion, to the effect that the relation of what we might call "model-

---

[1] Etchemendy [1990]

theoretic consequence" does not in general match up very well with the relation that underlies the correctness of ordinary inference, is in my view (which is apparently the minority one) essentially the correct one to draw. In any case, the book made it clear to me that the kinds of questions that had always fascinated me were, at least on occasion, susceptible to careful philosophical treatment.

The second thing that made it clear to me that the philosophy of logic could be done well was reading some history of logic, and discovering that the titans of modern logic were themselves philosophers at heart. I was intrigued to discover that Frege, Hilbert, Cantor, Zermelo, Brouwer, and Gödel, just to name a few, were deeply invested in purely-philosophical questions about the nature of logic, and about the role of formal systems in studying logic and mathematics. The debate between Frege and Hilbert over the nature of independence-proofs, to choose my own favorite example, was for me an eye-opening exchange. Here the subject-matter is that of the relationship between a purely-formal technique (essentially, the construction of models) and a handful of pre-formal questions (e.g. the independence of the parallels postulate, or the consistency of a theory). The debate involves not just the central issue of whether or not Hilbert's model-construction technique is a reliable method for establishing independence and consistency, but also a series of important related questions, having to do with the nature of mathematical axioms, the connection between formal deduction and intuitive provability, the importance of semantics to logic, and so forth. For me, debates of this kind between those involved in laying the foundations of logic were a vivid demonstration both of the importance of the debates themselves, and of the possibility of treating them with rigor and care.

The first issue that I thought I might make some headway on had to do with the connection between logic and mathematics, specifically the logicist attempts to "reduce" mathematics to logic. I soon discovered that there are as many different logicisms as there are logicists, with very different criteria of success, and different implications for the nature of the reduced and the reducing theories. Thinking that I might write a dissertation exploring those differences, the first chapter of which would have been on Frege, I soon found that Frege's logicism itself was sufficiently rich to keep me busy for quite a long time. The first chapter became the whole dissertation, as I have since learned is not an unusual pattern in dissertation-writing. My interests have since branched out to the history and philosophy of logic more broadly, but the issues raised in Frege's writings – see below – continue to play a central role in my work.

2. What are your main contributions to the philosophy of logic?

My work has mainly focused on Frege, and on a handful of philosophical issues that arise in his work. The technical details of Frege's work are relatively well known, but it has seemed to me that there's a lot that isn't yet well understood regarding the philosophical lessons to be learned from the successes and the failures of various parts of that work. Two of the issues I've been especially interested in are these: (i) Frege's conception of the connection between formal systems of logic and ordinary, non-formalized relations of entailment, consistency, independence, and other pretheoretic logical notions, and (ii) Frege's conception of the relation of "reducibility," that relation that he attempted to establish between logic and a large part of mathematics.

As to the first: Frege takes it that these pretheoretic relations all obtain between entities that he calls "thoughts," the things that are expressed by meaningful sentences. This includes, in a way that's surprising to a modern audience, the sentences of a formal system: Frege, unlike most of his successors, does not view the sentences of a formal deductive system as partly uninterpreted, but takes them, just like the sentences of natural languages, to express determinate thoughts. This makes it easy to characterize the relationship between *derivability* in a good formal system and *entailment*: a well-designed formal system is one that lets us *derive* a sentence S from a collection P of premise-sentences only if the thought expressed by S really is entailed by the thoughts expressed by the members of P. Frege did not, however, take it that the converse is generally to be expected: he held that the derivability of S from P is a sufficient, but not in general a necessary, condition for the logical entailment of S's thought by P's thoughts. Hence there is an in-principle divide between the important pretheoretic notion and the relation of formal derivability. This divide plays a large role in the debate mentioned above, between Frege and Hilbert over the provability of independence and consistency.[2]

Frege's view of the role of formal derivation is closely connected to the second issue, that of theoretical reduction. Here Frege's story is complicated, but the bottom line is that, as he conceives of it, a reduction of arithmetic to logic would have demonstrated, in a very straightforward way, that the true thoughts expressed by ordinary arithmetical sentences, e.g. the thought expressed by "Every natural number has a successor," are provable from fundamental truths of logic. The latter, the fundamental truths of logic, are themselves thoughts as well, and they are thoughts whose status as "logical" is, as Frege understands it, simply immediately evident.

---

[2] For my views on the Frege-Hilbert debate, see Blanchette [1996], [2007a], [2007b] and parts of [2012].

In my view, Frege's views and arguments about the nature of logic and related areas are generally powerful, and are essential to come to grips with if one wants to have clear views, oneself, about the nature of logic. Part of my contribution to the philosophy of logic, then, has been an extended treatment of Frege's conception of logic, and of his understanding of the connections between logic, language, and mathematics.[3] I argue that Frege's position (which I take to be often badly misunderstood) is an especially cogent one, with powerful and attractive implications for the nature of logic.

Another area in which I've worked concerns the relation of logical consequence, and the role of models in assessing that relation. I have argued, for example, that a fairly standard practice of treating models of formal theories as in some sense representatives of possible worlds, or of alternative states of affairs, is badly mistaken, and is the source of a good deal of confusion regarding the notion of model-theoretic entailment.[4] This work is closely connected to questions regarding the usefulness of second-order logic in assessing consequence, and to questions about the significance of the completeness of first-order logic.

3. What is the proper role of philosophy of logic in relation to other disciplines, and to other branches of philosophy?

Early in the 20th century, philosophical issues had a good deal to do with the design of formal systems and with the particular course followed in the development of modern logic. Questions of realism versus constructivism about the subject-matter of mathematics, questions about the relationship between truth and provability, and other clearly-philosophical issues explicitly motivated the work of the pioneers. In the last half-century, mainstream formal logic has grown into an independent discipline, in most areas a purely-mathematical one, so that to work in pure logic, one doesn't any longer need to be engaged with philosophical issues. Logic is in this way following the pattern of the physical sciences: initially motivated and shaped by philosophical issues and debates, the field has achieved a relatively-stable form, one that can, for the most part, be pursued independently of philosophical engagement.

The role of philosophy in relation to classical logic now, in the more-mature period (i.e. one in which a lot of logicians do no philosophy), is twofold. First of all, philosophers ask and answer questions having to do with the connection between logical systems in general and such pretheoretic notions as that of validity, truth, entailment, definition, analysis, form, number, measure, mathematical truth, and so on. Here

---

[3] See e.g. Blanchette [2012].

[4] Blanchette [2000], [2001].

the central questions straddle the boundary between the formal and the pretheoretic. We ask, for example, whether the validity, in a pretheoretic sense, of an ordinary argument can be reduced to, or analyzed in terms of, the formal validity of some logical system's representation of that argument; whether the representation of a theory's content via a categorical axiom-system is a particularly revealing method of representing it (and if so, why); whether consistency and independence in the formal logician's sense are reliable indicators of their pretheoretic namesakes; and so on. The hope is that we gain, through this kind of work, a better understanding both of what's going on with formal systems, and of the deeper, pre-formal notions themselves.

Secondly, philosophers ask, and answer, questions about the philosophical significance of various results in modern logic. What do we learn about the representative capacity of first-order logic when we discover that no first-order axiom-system with infinite models is categorical? (Does this provide pressure to move to a stronger language, or is categoricity not what we should be aiming for?) What do we learn about the deductive power of first-order logic when we learn that it's complete? (And what does the incompleteness of the stronger second-order logic tell us about its adequacy?) Does the fact that arithmetical truth is not recursively enumerable tell us something important about the nature of arithmetic? Does the independence of e.g. the continuum hypothesis from the typical axioms of set theory show us that there's something incoherent about some forms of set-theoretic realism? (Or that we should be searching for stronger axioms?) In general, logic itself provides a rich trove of precise formal results to which one needs to be sensitive in investigating a host of topics that have always been of central concern to philosophy; and it is the goal of the philosopher of logic to investigate exactly what those results imply with respect to those philosophical issues.

In addition to this engagement with classical logic, philosophers of logic currently play a fundamental role in the development and the discussion of alternative systems of logic. Modal logics, many-valued logics, free logics, relevance logics, paraconsistent logics, nonmonotonic logics, and so on are formal logics motivated by purely philosophical concerns. In general, the fundamental idea is that in some areas of discourse (e.g. discourse about necessity and possibility, or about empty domains, or about paradoxical situations), there are particular patterns, or rules, of inference that can be codified and studied via a formal logic specific to that kind of discourse. The role of the philosopher of logic here is to present reasons for adding to or departing from classical logic in these domains, to devise well-behaved formal systems reflecting the patterns in question, and to show that we can learn significant things about the kind of discourse in question, and about the knowledge involved in that discourse, by pursuing the formal project.

As to the relationship between the philosophy of logic and the rest of philosophy: here the connections are numerous. Modern philosophy of science is closely connected with the philosophy of logic via a fabric of interconnected questions and issues, and a shared history. The question of whether, and how, the axioms of a theory can provide implicit definitions of its central concepts has at times been a central question in modern philosophy of science, taking its defining concepts (those of modern axiom, implicit definition, and satisfaction) from 20th-century logic. The closely-related question of how we are to understand a scientific theory (whether for example as the deductive closure of its axioms, or as its class of models, or in some other way altogether) is informed by the logician's characterization of these notions. The questions of whether a scientific theory ought to be understood in terms of its "structure," and of how a theory applies to the observable world, are questions motivated in large part by the logician's characterization of theories in terms of structure. And so on. In general, the shared concern with the nature of axioms and theories, with the role of mathematics within theories, and with the application of theories to the real world, make the philosophy of logic and the philosophy of science natural partners.

Also extremely closely linked with the philosophy of logic is the philosophy of mathematics. This is partly because some of the central issues in each field are the same: in each case, knowledge in the field would seem to be obtained *a priori*, while the truths so obtained are essential to empirical research. This has made both logic and mathematics significant in their apparent challenge to various forms of empiricism, and has presented in each case an intriguing set of questions regarding the applicability of the discipline. The connection is also very close historically: modern formal logic was devised by those attempting to provide a foundation for mathematics, and by those who viewed it as a tool to be used within mathematical research. The question of how, exactly, logic and mathematics are connected – and in particular in what sense parts of mathematics are "reducible" to logic – was a driving question in a good deal of early logical research. That the reducibility question is of less-central importance today (though still alive in some quarters) is due largely to two factors. First, we have some negative results – e.g. the discovery that the most natural axiomatizations (Frege's, Russell's) that are rich enough to generate mathematics are either inconsistent or contain axioms whose logical status is questionable, and the lesson of Gödel's incompleteness theorem that no axiomatization will do the trick – that undermine the most straightforward reductionist projects. Secondly, via the development of set theory, the line between logic and mathematics has blurred: it is clear that set theory is sufficient for the "reduction" of much of mathematics, but set theory itself has enough arguably-mathematical content that this form of reduction is by no means the same project as the early logicist attempts to reduce mathematics to logic. The questions of what kinds of reduction are possible, and what we are to learn from them, both about the nature of mathematics and about the nature of logic, are on-going concerns in the philosophy of logic and the philosophy of mathematics.

The philosophy of language and epistemology too are closely linked with the philosophy of logic. As to the first: Logic is pursued by using particular highly-regimented languages, ones that no person has ever used in ordinary communication. Indeed, much of the time, the languages in question are only "partially interpreted," which is to say that, aside from the symbols for "and," "or," "not," "if-then", "for all", and so on, the symbols of the languages used to do logic have no fixed meanings. This gives rise to two issues having to do with what might roughly be called the "interpretation" of the languages in question. The first question is that of how the languages used in formal logic are related to natural languages. When we discover that a particular formula is provable in a good formal system from a set of formulas, what exactly does this tell us about the validity of an argument that's expressed in English? Here, the rough idea is that the formal argument is a "formalization of" an English-language argument, via the kinds of vague (indeed, they're *very* vague) translation-rules that one learns in a first logic course. What we want to know is, in part, what it is that a good formalization preserves from the original argument (do we intend to preserve just the references of the parts of the sentence? Or something more fine-grained, like a Fregean "sense" or an intension?), and – most importantly – whether what's preserved in this process is sufficiently robust to guarantee that the original argument and the formalized version share important logical properties. This raises the question of what, exactly, the bearers of the logical properties and relations are: can we understand validity and entailment to hold of series of sentences, i.e. bare strings of symbols? Or must we take these properties to hold of something more robust, like nonlinguistic propositions? Or is there some middle ground? The answer to this question, pursued by the philosopher of logic, is critical to the philosophy of language, since, for example, if one can only make sense of validity and entailment in terms of such extra-linguistic things as propositions, then it is incumbent on the philosopher of language to make room for such entities in the explanation of how language works. If on the other hand no sense can be made of the expression of such things in natural language, then (assuming that we can express valid arguments in ordinary language) the philosopher of logic must be able to make sense of the logical properties (validity, entailment) in terms of such things as sentences. This is just one instance of the rich connection between the philosophy of language and the philosophy of logic: in the end, the two are inseparable largely because we need languages in order to do logic, and we need to recognize logical relations in order to speak a language.

Epistemological questions were a driving force for some of the early pioneers of logic. Frege and Russell were both interested in the question of whether a "reduction" of mathematics to logic would help to explain the nature of mathematical knowledge and its certainty. Frege's student Carnap, along with a number of modern positivists, hoped that the close connections between mathematics and logic could be exploited to show that mathematics poses no problem for empiricism. Brouwer thought that the nature of mathematical knowledge placed strict limits on the reliability of various forms of logical inference. Gödel, holding a radically different picture of mathematical knowledge, held a correspondingly different view of the fundamental rules of logic. And so on. Logic has forever been viewed as a tool of *inference*, i.e. a tool by means of which we obtain new knowledge from old, with the result that views about the nature of knowledge and views about the laws of logic are inextricably connected. As above, epistemological concerns have been a driving force behind the development of a large number of alternative systems of logic, some of which supplement, and some of which depart from, ordinary classical logic. The connection between epistemology and the philosophy of logic is a robust two-way street: while the philosopher of logic must pay attention to the fact that logic is a tool for expanding knowledge, the epistemologist finds in logic a rich body of knowledge that cries out for explanation: is our knowledge of logic simply a species of linguistic or conventional knowledge, or is there something more robust, and harder to explain, behind our knowledge of logical truths?

In short, the philosophy of logic is deeply connected with a large proportion of the rest of philosophy, in no small part because logic itself is crucial to the reasoning that we engage in, and that we reflect upon, when we do philosophy.

4. What have been the most significant advances in the philosophy of logic?

Philosophy of logic, like most of philosophy, doesn't advance in discrete steps (like the proof of new earth-shaking theorems or the discovery of new species), but gains ground by slowly, carefully bringing some illumination to areas and questions that were earlier characterized by confusion.

My own view is that the most significant changes in this area over the last century have all had to do with increasing clarification about (i) the connections between language, mathematics and logic, and (ii) the nature of logical entailment itself.

On both of these issues, I think of Frege's work as having been especially significant. Frege understood that we need to have a clear appreciation of how language works if we are to even begin to understand the nature of logic. His idea, for example, that the principles of logic have to do not with expressions but with what they express, makes sense only against the background of a theory of meaning that takes something like sense, or intension, into account. And though he was not unique in this way, his clear and dogged pursuit of the connections between a theory of meaning and an account of the nature of logic served to set the agenda for much of modern philosophy. We now know that there are alternative ways of understanding meaning that will make sense of the connection between meaning and inference, and that there are arguments both for and against the specifically Fregean accounts of meaning and of the logical relations. But one enduring lesson of Frege's work is that theories of logical entailment and of meaning cannot be pursued in isolation from one another: whatever and however it is that our words and sentences mean, it has become clear that semantic properties are determined by, and determine, logical properties: our sentences mean what they mean in large part because of what they logically imply (and are implied by), and these relations of implication are determined in large part by what those sentences mean.

Frege's view that we can come to understand the nature of arithmetic by reducing it to logic, and that this reductionist project must involve both a careful axiomatization of the logic and careful conceptual analyses of arithmetical notions, has borne a great deal of fruit. We have learned that a certain conceptual richness of arithmetic, together with a lack of corresponding richness in pure logic, makes a reduction of the kind Frege envisioned impossible. This is as clear as philosophy ever gets, it seems to me, to establishing a firm and foundational result of a surprising kind. As a result, we have achieved a considerably clearer conception of what is at stake in the attempt to explain mathematical truth and mathematical knowledge, and also a considerably clearer conception of the bounds of the purely logical.

In the post-Fregean era, logic itself has developed dramatically, and some of the most important work in the philosophy of logic has involved the exploration of exactly what we learn from modern results in pure logic. Over the last eighty years or so, we have, for example, achieved a much clearer conception of the bounds of provability. We have learned that axiomatizations even of such apparently-simple theories as that of natural-number arithmetic are very hard to come by, and so we've recognized that the principles of inference needed for such theories are vastly richer than they at first appeared to be. We have learned that the notions of "true sentence" and of "sentence provable-in-system-S" are much less straightforwardly characterized than one might have thought, and that the difficulties in so characterizing them help us establish boundaries to the expressive power of formal systems. We've managed to characterize the notion of "computable function," and to show how surprisingly narrow the bounds of computability are. These and related modern results regarding the expressive and proof-theoretic limitations of formal systems seem to me to be extremely important in coming to understand how logic is related to language, to mathematics, and to reasoning in general.

5. What are the most important open problems in philosophy of logic, and what are the prospects for progress?

I don't have anything like a "top-ten" favorite list of discrete open problems, in part because most of the interesting questions in the philosophy of logic are so closely tied to one another as to form one large mass, rather than a handful of discrete problems. The connected questions here include those of the right way to understand alternative logics (where such logics conflict, is there one "right" logic, or can we make sense of different logics applicable to different projects or domains?), the significance of various connections between deductive and model-theoretic consequence relations, and hence of completeness and incompleteness results, and that of what, exactly, we should infer from the remarkable success of axiomatic set theory as a foundational tool both in logic and in mathematics. We also need to come to grips with the claims of competing foundational theories, especially that of category theory. In these areas, as in all parts of the philosophy of logic, it seems to me that it will be important for us to do a better job of understanding not just current formal developments, but also the history of the discipline.

One of the further issues about which I would like us to gain greater clarity is that of the nature of logical truth and logical entailment. There was enthusiasm in the early 20th century for the idea that logic is a matter, somehow, of convention, or a byproduct of the forms of our linguistic or mental representation. The attractiveness of such a view is relatively clear: if true, then we seem to have the outline of an answer to the questions of how we know the truths of logic, and of why the relation of logical entailment is immediately and universally applicable. But no clear statement of this idea seems plausible: as Quine has pointed out, it's very difficult to cash out a clear sense of "convention" in which logic can reasonably be said to hold "in virtue of convention;"

and it's not at all clear that there is a way of understanding "forms of representation" which fares any better.[5] One of the difficulties is that when we say that p holds "in virtue of" q, we often mean simply that p is logically entailed by q; and when the p in question is logic itself, such an account is a non-starter. Formulating a clear statement of what exactly the principles of logic *are* (are they rules about representation? About truth? And what's a *rule* anyway?), or a clear line of reasoning to the effect that this is a misguided question, seems to me to be an issue that's still quite unsettled, and worth pursuing. As to prospects for progress: I'm optimistic. We have come a long way in understanding the nature of logic over the last century or so, and I see no reason that we can't press on quite a bit further.

Bibliography

Blanchette, P. [1996] "Frege and Hilbert on Consistency," *Journal of Philosophy* XCIII, 317-336.

Blanchette, P. [2000] "Models and Modality," *Synthese* Vol 124, No. 1 / 2, 45-72.

Blanchette, P. [2001] "Logical Consequence," *Blackwell Guide to Philosophical Logic* (Blackwell: Malden MA and Oxford), 115-135

Blanchette, P. [2007a] "The Frege-Hilbert Controversy," *Stanford Encyclopedia of Philosophy* http://plato.stanford.edu/entries/frege-hilbert/

Blanchette, P. [2007b] "Frege on Consistency and Conceptual Analysis" *Philosophia Mathematica* 15, 3 (2007) pp 321-346

Blanchette, P. [2012] *Frege's Conception of Logic*, Oxford University Press

Etchemendy, J. [1990] *The Concept of Logical Consequence*, Harvard University Press, Cambridge MA.

Quine, W. V. O. [1936] "Truth By Convention," reprinted in *The Ways of Paradox and Other Essays*, Harvard University Press 1966.

---

[5] Quine [1936]

5

Otávio Bueno

Professor of Philosophy
University of Miami, USA

---

1. Why were you initially drawn to the philosophy of logic?

My interest in the philosophy of logic started when I attended, as an undergraduate student, the seminars of Professor Newton da Costa at the University of São Paulo. He made it very clear that careful philosophical reflection about logic cannot be done without proper attention to the many connections that logic bears not only to mathematics and the sciences, but also to ontology and epistemology. Very early on, it was clear to me that the philosophy of logic, properly done, goes hand in hand with central areas of philosophy, and that was a significant attraction for me. (The philosophy of logic is, of course, not unique in this respect. But this point is particularly striking about it.)

The philosophy of logic also allows one to explore extremely sharp and well-defined problems — several of which can be addressed formally — possessing a significant impact for larger areas of philosophy. It is intriguing how much can be learned about a certain sub-field by formulating some problems formally and then examining the significant impact their solutions have for the broader philosophical landscape. A classical example in logic (although, of course, it has huge consequences for the philosophy of logic) is Gödel's incompleteness theorems. By formalizing the concept of provability, quite unexpected results regarding the incompleteness of arithmetic and the unprovability of consistency emerged — results that challenged deeply held assumptions about the philosophy and the foundations of mathematics.

The connections that the philosophy of logic has with other sub-fields of philosophy allowed me to explore, over many years, the issues it raises with collaborators working in logic, philosophy of science, philosophy of mathematics, and metaphysics. It has been, in fact, a privilege to collaborate with a number of terrific people on these issues.

Several motivations for the development of my views in the philosophy of logic emerged from discussions with Newton da Costa, who has a remarkable vision for so many significant issues in philosophy. I have written many works with him, developing the details of the resulting views, ranging from non-classical logics (in particular, paraconsistent and non-reflexive logics) to quasi-truth and partial structures and their significance (see, e.g., da Costa and Bueno [2001], [2007], and [2009]).

Similarly important has been my collaboration with Steven French, with whom I developed various aspects of the use of logic for the understanding of salient features of scientific practice (particularly in terms of partial structures — another great idea of Newton da Costa's — see, e.g., da Costa, Bueno, and French [1998a], Bueno and French [2011], and [2012]). With Scott Shalkowski I have explored the role that primitive modality plays in the proper understanding of logic, logical constants, and logical consequence. We have been developing a modalist approach to central areas of the philosophy of logic, and examining their implications for the philosophy of mathematics as well as the metaphysics and the epistemology of modality (see Bueno and Shalkowski [2000], [2004], [2009], and [2013]).

With Mark Colyvan, I have developed a framework to represent the application of mathematics (which we call the *inferential approach*), and which has significant implications for broader issues of scientific representation (Bueno and Colyvan [2011]), together with problems related to the revision of logical principles (Bueno and Colyvan [2004]) and semantic paradoxes (Bueno and Colyvan [2003a] and [2003b]).

With Décio Krause, I have explored the role played by logic in the proper understanding of scientific theories and their models (Krause and Bueno [2007] and [2010], and da Costa, Krause and Bueno [2010]) together with its significance for an account of scientific reasoning (da Costa, Krause, and Bueno [2007]). Many of these issues have been developed in collaboration with Newton da Costa. We thus reach a full circle.

2. What are your main contributions to the philosophy of logic?

My main contributions to the philosophy of logic, often developed in collaboration with those mentioned above, involve: (a) the defense of logical pluralism, (b) the emphasis on the significance of non-classical logics (particularly, paraconsistent logics) for the philosophical understanding of logic, and (c) the defense of logical non-apriorism (that is, the view that logical principles can be revised on non-a priori grounds). I will consider each of these points in turn.

(a) *Logical pluralism* is the view according to which there is a plurality of logics depending on the domain one considers, and, in fact, typically there are several logics appropriate to a given domain (see da Costa and Bueno [2001], Bueno [2002a], Bueno [2011a], Bueno and Shalkowski [2009] and [2013]). For example, to capture constructive features of mathematical reasoning, classical or paraconsistent logics are clearly inadequate, but intuitionistic logics are not. To accommodate inconsistent bits of information without triviality (that is, without deriving everything as well), intuitionist and classical logics are inadequate, but paraconsistent logics are not. To secure the strongest possible logic (with regard to the consequences that are obtained), paraconsistent and intuitionistic logics are not adequate, but classical logic is. (Of course, some non-classical logicians will typically question whether some such consequences are indeed valid. But that is part of the ongoing debate.)

In other words, the logical pluralist insists that there are several logics across several domains and multiple ones in a single domain too. For example, it is clear from Newton da Costa's hierarchy of paraconsistent logics (the C-logics; see da Costa [1974]) that if a given particular paraconsistent logic is adequate for a particular domain, there are several others that are just as adequate. This does not mean, of course, that any logic is adequate for any domain. Logical pluralism is not logical relativism. Some logics are just inadequate for certain domains, that is, certain areas of inquiry (as the examples mentioned in the paragraph above illustrate).

(b) The *philosophical significance of non-classical logics* becomes clear in this context: these logics challenge the adequacy of certain assumptions made by classical logic and provide an alternative understanding of certain features of logic, such as the nature of logical constants (Bueno and Shalkowski [2013]), the scope of logical principles (da Costa, Bueno, and French [1998*b*], da Costa and Bueno [2001], da Costa, Krause, and Bueno [2007], and Bueno [2010*a*]), the interpretation of quantifiers (da Costa and Bueno [2009]), as well as the role of models and primitive modality in the proper formulation of logics (Bueno and Shalkowski [2009] and [2013]).

(c) The *possibility of revising logical principles on non-a priori* grounds is also a significant aspect of my work in the philosophy of logic (da Costa and Bueno [2001], Bueno and Colyvan [2004], and Bueno [2010*b*]). Although logical principles are accepted on a priori grounds, they can be revised due to non-a priori — and, in some instances, even empirical — considerations. The crucial point is that for the acceptance of a logical principle, no empirical considerations are needed. It is ultimately a matter of considering the relevant logical form. (Note that I am talking about *acceptance* rather than full justification. It is arguably unclear how to establish any such unquestionable justification.) However, when applied to a particular domain (e.g. to describe some specified objects and their relations), the principles or inferences in question can yield the wrong results. In this case, a revision of such principles or inferences is in order. For example, empirical considerations lead to a revision of the distributivity law in the foundations of quantum mechanics (see Putnam [1979], da Costa and Bueno [2001], and Bueno and Colyvan [2004]).

It may be objected that we need some logical principles to adjudicate between other logical principles. In this case, how is it possible to revise the contentious principles without somehow begging the question? In response, the central idea is that we can always adopt, pragmatically, some other logical principles to assess the consequences of the controversial ones. We need not assume, say, excluded middle while reasoning about what follows from its adoption. And through these considerations, it is eventually possible to assess the principles in question, and as a result, to revise them.

It should be clear now that logical non-apriorism, logical pluralism, and the philosophical significance of non-classical logics all go hand in hand. They are mutually supporting views about logic and provide a unified approach to philosophical problems in the field.

3. What is the proper role of philosophy of logic in relation to other disciplines, and to other branches of philosophy?

The philosophy of logic, as I mentioned above, has important connections to other branches of philosophy, such as: (i) metaphysics, (ii) epistemology, (iii) philosophy of science, and (iv) philosophy of mathematics. I will consider them in turn.

(i) *Careful philosophical reflection about the nature of quantification cannot be done without attention to the underlying metaphysics.* For instance, it seems that classical quantification presupposes the identity of the objects that are quantified over. However, non-reflexive logics, which do not assume that identity applies to every object in their domains, have no such presupposition. This raises the possibility of understanding the relevant quantifiers without presupposing the identity of the objects in question (see da Costa and Bueno [2009]). Moreover, an adequate specification of the scope of logical principles requires a clear understanding of modality (Bueno and Shalkowski [2009]), and the same goes for the characterization of logical constants (Bueno and Shalkowski [2013]). After all, if certain possibilities are allowed, such as inconsistent situations, then certain principles of classical logic are violated (such as *explosion*, according to which everything follows from a contradiction). As a result, the scope of these principles needs to be restricted, and different logical constants can then be introduced.

(ii) *An important goal of philosophical reflection about logic is an account of our knowledge of logical principles.* Clearly, this cannot be done without attention to *epistemology*, and by engaging, in particular, with accounts of a priori knowledge (Bueno [2011b]). Some views of the epistemology of logic require metaphysical assumptions about the nature of the objects under consideration (propositions, facts, universals). This poses a particular difficulty for the resulting epistemological views, given the need for explaining how the knowledge of the relevant objects is possible. Other views do not have such metaphysical commitments. They need then to explain what logical knowledge is about, and distinguish those who have a lot of it from those who do not (see Field [1989], and Casullo [2003]).

(iii) *Philosophical views about logic also bear important connections with the philosophy of science*. To make sense of scientific reasoning and different styles of reasoning within scientific activity, a proper understanding of logic seems crucial (see Bueno [2012b]). After all, depending on the logic one considers, different styles of reasoning may emerge. For instance, constructive and non-constructive styles of reasoning are importantly different, and these differences emerge, in part, from the different conceptions of logic at issue. Similarly, to understand the various roles that models play in scientific practice, including their heuristic role in theory construction and theory development, an account of the philosophy of logic is also significant (see Bueno, French, and Ladyman [2012a] and [2012b], da Costa and French [2003], and da Costa and Bueno [2009]). After all, which models can be formulated depends on the particular framework that is used to characterize them, and the expressive resources of the framework, in turn, depend on its underlying logic. But the choice of the logic in question is made, in part, on the basis of its philosophical understanding and the overall resources it provides.

Another important aspect of how philosophy of logic is connected with the philosophy of science is in the understanding of the relationship between mathematical and empirical features in the explanation of phenomena (Bueno and French [2011] and [2012], Bueno, French and Ladyman [2002], and Bueno [1997]). A class of such explanations involves transferring of structure from one domain to another. The resources available for implementing such transfers also depend on the framework that is used to formulate the relevant models, which, as just noted, also depends on the underlying logic that is adopted. Typically, in science the underlying logic is classical. But since in some cases non-classical logics can be used, the issue of changing the underlying logic becomes relevant. More generally, understanding the process of theory change and belief change, including the role of inconsistencies in this process, also requires a proper account of the underlying logic and its nature. These are, of course, philosophical issues about logic that also bear on general problems in the philosophy of science (see da Costa and Bueno [1998], [2001], and [2007], and Bueno [2000], [2002b], and [2006]).

(iv) Finally, *philosophical conceptions about logic often go hand in hand with views in the philosophy of mathematics* (see Bueno [2012a], [2005], [2001], and [1999]). Logicism is, of course, an obvious example, given its attempt to reduce arithmetic to logic (see Frege [1884/1950], Hale and Wright [2001], and Bueno [2001]). Those who favor second-order logic can also argue that, properly interpreted in terms of plural quantifiers (Boolos [1998]), and given its expressive resources, second-order logic can be used to provide a defense of nominalist views that avoid commitment to abstract objects in mathematics (Field [1980], Bueno [2010a], and Bueno [2009]). The logic is made stronger (expressively), but the ontological commitments are reduced.

4. What have been the most significant advances in the philosophy of logic?

There have been a number of significant advances in the philosophy of logic. I will mention a few of them.

(i) *The development of plural interpretations of second-order quantifiers and their applications* (Boolos [1998], Lewis [1991], and Linnebo [2003]). This interpretation of monadic second-order logic revived the interest in philosophical discussions of second-order logic, and has implications for discussions of nominalism and platonism in the philosophy of mathematics, the interpretation of set theory, and the relation between natural language and formalized theories (for a helpful survey, see Linnebo [2012]).

(ii) *Challenges to the model-theoretic understanding of logical consequence and responses to these challenges* (Etchemendy [1990], Sher [1996], Gomez-Torrente [1996], and Bueno and Shalkowski [2013]). The model-theoretic approach to logical consequence is, arguably, the most widely accepted view in logic and in its philosophy. Challenges to it have identified hidden assumptions and difficulties. Not surprisingly, the responses have attempted to offer robust answers to the troubles that were raised. Even if the original objections may not go through, there is something fundamentally right about the challenge.

(iii) *The defense of certain forms of nominalism about logical formalism* (Azzouni [2006]). This is a significant difficulty for any nominalist view, and needs to be properly addressed by any fully developed form of nominalism (both in mathematics and in logic): how can nominalists make sense, nominalistically, of the very formalism they use? As I will note below, although significant advances have been made in this area, there are still significant issues to be addressed (needless to say, this is true of virtually any noteworthy philosophical problem!).

(iv) *The development of different kinds of logical pluralism* (da Costa and Bueno [2001], Beall and Restall [2006], and Bueno and Shalkowski [2009]). One of the most extraordinary facts about logic in the last century is the plurality of logical systems that have been developed. A careful philosophical reflection about this fact and its significance is crucial to a proper understanding of logic and its nature. Interestingly, there are different kinds of logical pluralism: some emphasize cases, others stress domains; some incorporate primitive modality; others do not.

5. What are the most important open problems in philosophy of logic, and what are the prospects for progress?

There are several important open problems in philosophy of logic. I will mention two.

(a) *A proper understanding of logic requires a metaphysics*. But which metaphysics is a significant open problem. Many influential views about the foundations of logic incorporate metaphysical commitments of a problematic sort: abstract objects, universals, facts, and propositions are just a few examples of such commitments. They require a proper account of the epistemology of such objects, one that goes hand in hand with an account of our mechanisms of access to the relevant entities. These views are Platonist in the sense that they posit non-spatiotemporal, causally inert entities, whose knowledge ultimately requires explanation. Platonist views about logic try to address this issue, but with only some degree of success. After all, in the end, it is unclear that they really succeed in establishing knowledge of the relevant objects. Nominalist views similarly attempt to provide a solution. But they still face difficulties. For it is unclear that the account of logic consequence they provide ultimately works. A fundamentally different approach seems required to move us beyond such stalemate.

(b) Another significant open problem is the *understanding of the role played by logic in styles of reasoning used in the sciences and in mathematics*. This question is partially conceptual, partially empirical. It is conceptual in that it requires a proper characterization of the variety of such styles and their connections with particular logics (when appropriate). In some cases, the connection is clear enough. For example, constructivism — as an approach to the foundations of mathematics and as a philosophical view about the nature of mathematics and logic — provides a distinctive style of reasoning. Not surprisingly, its associated logic is intuitionist. In other cases, however, the situation is more complex. After all, the same logic may be associated with different styles of reasoning. The difficulty then consists in determining which logic (if any) is being used, and to provide a proper philosophical account of the situation.

But the problem of understanding the role played by logic in styles of reasoning is also empirical in that it depends on particular traits of actual scientific and mathematical practice, and the proper study of these traits. Any such study will involve a logic and, thus, considerations regarding the choice of logical principles and how to settle foundational debates about logic will need to be invoked as well. This is philosophy, after all. We are always at sea.

Bibliography

Azzouni, J. [2006]: *Tracking Reason: Proof, Consequence, and Truth*. New York: Oxford University Press.

Beall, JC, and Restall, G. [2006]: *Logical Pluralism*. Oxford: Clarendon Press.

Boolos, G. [1998]: *Logic, Logic, and Logic*. Cambridge, Mass.: Harvard University Press.

Bueno, O. [1997]: "Empirical Adequacy: A Partial Structures Approach", *Studies in History and Philosophy of Science 28*, pp. 585-610.

Bueno, O. [1999]: "Empiricism, Conservativeness and Quasi-Truth", *Philosophy of Science 66*, pp. S474-S485.

Bueno, O. [2000]: "Empiricism, Mathematical Change and Scientific Change", *Studies in History and Philosophy of Science 31*, pp. 269-296.

Bueno, O. [2001]: "Logicism Revisited", *Principia 5*, pp. 99-124.

Bueno, O. [2002a]: "Can a Paraconsistent Theorist be a Logical Monist?", in W. Carnielli, M. Coniglio, and I. D'Ottaviano (eds.), *Paraconsistency: The Logical Way to the Inconsistent* (New York: Marcel Dekker), pp. 535-552.

Bueno, O. [2002b]: "Mathematical Change and Inconsistency: A Partial Structures Approach", in Joke Meheus (ed.), *Inconsistency in Science* (Dordrecht: Kluwer Academic Publishers), pp. 59-79.

Bueno, O. [2005]: "On the Referential Indeterminacy of Logical and Mathematical Concepts", *Journal of Philosophical Logic 34*, pp. 65-79.

Bueno, O. [2006]: "Why Inconsistency Is Not Hell: Making Room for Inconsistency in Science", in Erik Olsson (ed.), *Knowledge and Inquiry: Essays on the Pragmatism of Issac Levi* (Cambridge: Cambridge University Press), pp. 70-86.

Bueno, O. [2009]: "Mathematical Fictionalism", in Otávio Bueno and Øystein Linnebo (eds.), *New Waves in Philosophy of Mathematics* (Hampshire: Palgrave MacMillan, 2009), pp. 59-79.

Bueno, O. [2010a]: "A Defense of Second-order Logic", *Axiomathes 20*, pp. 365-383.

Bueno, O. [2010*b*]: "Is Logic A Priori?", *Harvard Review of Philosophy 17*, pp. 105-117.

Bueno, O. [2011*a*]: "Relativism in Set Theory and Mathematics", in Steven Hales (ed.), *A Companion to Relativism* (Oxford: Blackwell), pp. 553-568.

Bueno, O. [2011*b*]: "Logical and Mathematical Knowledge", in Sven Bernecker and Duncan Pritchard (eds.), *Routledge Companion to Epistemology* (London: Routledge), pp. 358-368.

Bueno, O. [2012*a*]: "An Easy Road to Nominalism", *Mind 121*, pp. 967-982.

Bueno, O. [2012*b*]: "Styles of Reasoning: A Pluralist View", *Studies in History and Philosophy of Science 43*, pp. 657-665.

Bueno, O., and Colyvan, M. [2003*a*]: "Yablo's Paradox and Referring to Infinite Objects", *Australasian Journal of Philosophy 81*, pp. 402-412.

Bueno, O., and Colyvan, M. [2003*b*]: "Paradox without Satisfaction", *Analysis 63*, pp. 152-156.

Bueno, O., and Colyvan, M. [2004]: "Logical Non-Apriorism and the 'Law' of Non-Contradiction", in Graham Priest, JC Beall, and Brad Armour-Garb (eds.), *The Law of Non-Contradiction: New Philosophical Essays* (Oxford: Clarendon Press), pp. 156-175.

Bueno, O., and Colyvan, M. [2011]: "An Inferential Conception of the Application of Mathematics", *Noûs 45*, pp. 345-374.

Bueno, O., and French, S. [2011]: "How Theories Represent", *British Journal for the Philosophy of Science 62*, pp. 857-894.

Bueno, O., and French, S. [2012]: "Can Mathematics Explain Physical Phenomena?", *British Journal for the Philosophy of Science 63*, pp. 85-113.

Bueno, O., French, S, and Ladyman, J. [2012*a*]: "Empirical Factors and Structure Transference: Returning to the London Account", *Studies in History and Philosophy of Modern Physics 43*, pp. 95-104.

Bueno, O., French, S, and Ladyman, J. [2012*b*]: "Models and Structures: Phenomenological and Partial", *Studies in History and Philosophy of Modern Physics 43*, pp. 43-46.

Bueno, O., French, S, and Ladyman, J. [2002]: "On Representing the Relationship between the Mathematical and the Empirical", *Philosophy of Science 69*, pp. 497-518.

Bueno, O., and Shalkowski, S. [2000]: "A Plea for a Modal Realist Epistemology", *Acta Analytica 15*, pp. 175-193.

Bueno, O., and Shalkowski, S. [2004]: "Modal Realism and Modal

Epistemology: A Huge Gap", in Erik Weber and Tim De Mey (eds.), *Modal Epistemology* (Brussels: Royal Flemish Academy of Belgium), pp. 93-106.

Bueno, O., and Shalkowski, S. [2009]: "Modalism and Logical Pluralism", *Mind 118*, pp. 295-321.

Bueno, O., and Shalkowski, S. [2013]: "Logical Constants: A Modalist Approach", *Noûs 47*, pp. 1-24.

Casullo, A. [2003]: *A Priori Justification*. New York: Oxford University Press.

da Costa, N.C.A. [1974]: "On the Theory of Inconsistent Formal Systems", *Notre Dame Journal of Formal Logic 15*, pp. 497-510.

da Costa, N.C.A., and Bueno, O. [1998]: "Belief Change and Inconsistency", *Logique et Analyse 161-162-163*, pp. 31-56.

da Costa, N.C.A., and Bueno, O. [2001]: "Paraconsistency: Towards a Tentative Interpretation", *Theoria 16*, pp. 119-145.

da Costa, N.C.A., and Bueno, O. [2007]: "Quasi-Truth, Paraconsistency, and the Foundations of Science", *Synthese 154*, 2007, pp. 383-399.

da Costa, N.C.A., and Bueno, O. [2009]: "Non-Reflexive Logics", *Revista Brasileira de Filosofia 232*, pp. 181-196.

da Costa, N.C.A., Bueno, O., and French, S. [1998*a*]: "The Logic of Pragmatic Truth", *Journal of Philosophical Logic 27*, pp. 603-620.

da Costa, N.C.A., Bueno, O., and French, S. [1998*b*]: "Is there a Zande Logic?", *History and Philosophy of Logic 19*, pp. 41-54.

da Costa, N.C.A., and French, S. [2003]: *Science and Partial Truth*. New York: Oxford University Press.

da Costa, N.C.A., Krause, D., and Bueno, O. [2007]: "Paraconsistent Logics and Paraconsistency", in Dale Jacquette (ed.), *Philosophy of Logic* (Amsterdam: North-Holland), pp. 791-911.

da Costa, N.C.A., Krause, D., and Bueno, O. [2010]: "Issues in the Foundations of Science, I: Languages, Structures, and Models", *Manuscrito 33*, pp. 123-141.

Etchemendy, J. [1990]: *The Concept of Logical Consequence*. Cambridge, MA: Harvard University Press.

Field, H. [1980]: *Science without Numbers: A Defense of Nominalism*. Princeton, N.J.: Princeton University Press.

Field, H. [1989]: *Realism, Mathematics and Modality*. Oxford: Basil Blackwell.

Frege, G. [1884/1950]: *The Foundations of Arithmetic*. (English translation by J.L. Austin.) Oxford: Blackwell.

Gomez-Torrente, M. [1996]: "Tarski on Logical Consequence", *Notre Dame Journal of Formal Logic 37*, pp. 125-151.

Hale, B., and Wright, C. [2001]: *The Reason's Proper Study: Essays Towards a Neo-Fregean Philosophy of Mathematics*. Oxford: Clarendon Press.

Krause, D., and Bueno, O. [2007]: "Scientific Theories, Models, and the Semantic Approach", *Principia 11*, pp. 187-201.

Krause, D., and Bueno, O. [2010]: "Ontological Issues in Quantum Theory", *Manuscrito 33*, pp. 269-283.

Lewis, D. [1991]: *Parts of Classes*. Oxford: Blackwell.

Linnebo, Ø. [2003]: "Plural Quantification Exposed", *Noûs 37*, pp. 71-92.

Linnebo, Ø. [2012]: "Plural Quantification", in Edward N. Zalta (ed.), *The Stanford Encyclopedia of Philosophy*. (Spring 2013 edition.) URL = <http://plato.stanford.edu/archives/spr2013/entries/plural-quant/>.

Putnam, H. [1979]: *Mathematics, Matter and Method*. Philosophical Papers, volume 1. (Second edition.) Cambridge: Cambridge University Press.

Sher, G. [1996]: "Did Tarski Commit Tarski's Fallacy?", *Journal of Symbolic Logic 61*, pp. 653-686.

6

James Cargile

Professor
Corcoran Department of Philosophy, University of Virginia

Logic puzzles will be recognized as a distinctive kind without the introduction of the title, like (perhaps including) math puzzles but not numerical or geometric. The features of validity and necessity are not restricted as to subject matter. I learned to call assessing puzzling arguments "Logic" and felt that was the subject for me, from an early age. The title "Philosophy of Logic" has never had such an association for me, but thinking philosophically about Logic is part of Logic.

One memorable early puzzle was roughly like this: 1. If the sentence A: "Sentence A is not true", is true, then things are as it says, so A is not true. 2. Therefore, the assumption A is true entails its negation, so by *Reductio*, A is not true. 3. But A says that A is not true, so, if A is not true, then things are as it says, so that A is true. 4. Therefore, A is both true and not true. Some find this nonsensical, some find it amusing and some nowadays take it as a proof that some sentences are both true and not true. To me, it was immediately obvious that the argument could not be correct, because it is a *Truth* (of Logic) that nothing whatever could be both true and not true (at the same time and in the same respect). But the premises and the inferences were plausible. That they entail a falsehood proves them wrong, but does not explain what is wrong.

The commonest answer offered was rejection of self reference. A sentence cannot say, about itself, that it is not true, or that it is true. A sentence cannot refer to itself at all: (i) If a sentence x is about a sentence y, then x is somehow of a higher level than y, so it is impossible that x=y. (i) was unsatisfactory. If it were about itself it would be self-violating and if not about itself, it would lack the generality it pretends to. (ii) No sentence is both true and false (at the same time and same respect). (ii) expresses a truth about all sentences, including itself, contrary to the rejection. And (iii) "This sentence has five words" can obviously be used to tell us a plain truth about itself. It could be used to say something about a different sentence, but that is not a necessary restriction on its use.

Gilbert Ryle offered an answer for a similar puzzle, with his notable clarity and simplicity. The version involved (B) "The current statement is false". Ryle observes that the question "Namely, what statement?" leads to a viciously regressive series of answers, "The statement, namely, the statement..." He says "...no statement of which we can even ask whether it is true or false is ever adduced". This seems, initially, a fair reason for setting aside B as vacuous. Its application to A is less clear, since "This sentence, namely..." can be non-regressively answered by merely citing the sentence A. Ryle was working in terms of statements as what is said (if anything) by a person, rather than sentences (like A) taken by themselves. There is a profound difference. Someone who says to us "What I am saying is false" deserves to be asked just what it is he is saying, while asking a sentence itself might seem absurd. I adopted a crucial pretense--- that the sentence is talking to us, even though it may be a very unsatisfactory interlocutor (see Phaedrus 275e).

If (iii) is telling us it has five words, there is no regress problem. The difference is that (iii) is about itself as a sentence, while A has to be about more than that – what it, the sentence is saying. Reporting what is said is a logically fundamental activity which is badly misunderstood if taken as accomplishable merely by repeating the words used or mechanical grammatical variations. Repetition has finality. Finding what is said can be lengthy or indefinite. One start is that A is saying "A either says nothing at all or what it says is not true". That leads into a Rylean regress and allows his verdict. However, Ryle's approach does not apply well to (C) "Every Cretan statement is false". "Namely?" is not an appropriate question for a general, as opposed to, an individuating, reference. The approach can be repaired, switching to "for example?" ("e.g. riders"). They lead to vicious regress using the words of C, which (from a Cretan) are eligible as an example. But if that sets aside C, it equally sets aside (ii) and other general logical laws. Ryle's answer is another way of banning self reference, with the same liabilities.

What struck me about Rylean expansions of A, B or C, was not the regressive quality, but rather, their inconsistency. A verbal claim of falsity or non-truth at one stage would itself be called false or non-true at a later stage. No one ever worried about D: "The current statement is both true and not true". There is as much a Rylean regress as with B. You could, of course, follow Ryle's verdict. But it is notable that the inconsistency of D is good reason to call it false.

With B, offered by a person, if we ask "Namely, what statement?", the reply "Just that!" is a poor dialectical performance. It may be replied that the speaker is presenting us with a poser, a challenge to our logic. Very well, but he has to concede that the saying which he is telling us is false, is his saying that what he is telling us is false. He poses as telling us something and as telling us that whatever it is, is false. He will, typically, insist those are the same thing. But saying something and saying it is false, is self contradictory. That is fair grounds for judging his performance false. We say it is false. He insists he agrees with us. But we are not engaged in making his statement, only in judging it. And we judge the expansion inconsistent. He can do that too, but he is also the source of the claim to be expanded. We are not.

That is what occurred to me in meditating on Ryle's essay, and I found it liberating. It was also excellent for C. In citing examples of Cretan statements called false, C itself will come up, if the speaker is a Cretan. This does not require his knowing he is a Cretan or having any sophistical intention. He would be asserting, albeit unknowingly, that it is false that all Cretan statements are false. That is inconsistent with asserting that all Cretan statements are false. So a Cretan assertion of C is false, regardless of the truth record of all other Cretan claims.

These answers did not initially appeal to general logical rules. They suggested some: (I) To assert a proposition is one and the same as to predicate truth of it and to deny a proposition is the same as predicating nontruth, which in application to a proposition is the same thing as falsity. (II) To assert that all Xs are Ys is to predicate being a Y of every X. (III) To assert that if P then Q and that P is to assert that Q. The asserting and predicating in these rules is not knowingly asserting or predicating. Anyone who asserts that all men are mortal, attributes mortality to me, even though he may have no notion of me. Rule II allows an important strengthening of III (taking "If P then Q" as "Every P case is a Q case"): asserting that if P then Q, in the case in which P is true, is to assert Q, whether or not you asserted P. (This is taking asserting that if P then Q to be conditionally asserting Q. An objection is that one might assert that P materially implies Q, simply on the grounds that not-P, without making a conditional assertion of Q. That confuses limited reasons for offering a guarantee with a limited guarantee.)

These rules about asserting do not apply to believing, fearing, hoping, knowing, and other notions, factive or otherwise. A Cretan who believes that every Cretan belief is false would have a very peculiar belief, hard to make sense of, but it would not require that he believes it false that every Cretan belief is false. To make sense of a sincere belief here would require that he believe he is not a Cretan, as a part of the belief he expresses with "Every Cretan belief is false". This is a difficult point to make clear and crucially depends on refusing to make the units of logical analysis for these cases, sentences, as opposed to what sense can be made of sentences as used by a particular person, in good dialogue with that person. We can ask what a sentence says, but not what it believes.

Furthermore, these rules did not suffice for paradoxes such as one in which the policeman says "Something the prisoner deposes is false" and the prisoner deposes only "Everything the policeman deposes is true". Nor did they answer Grelling's "Heterological", where that means being a (linguistic) predicate which expresses a property of which it is not an instance. My answer to the latter is that "heterological" is not heterological, because it does not express a property at all. That was not an unusual answer, since rejecting self reference encourages it, but the explanation was difficult for me, having no objection to self reference. One can say, of a predicate whose meaning he does not know, that it is heterological, that it expresses some property (which he cannot identify), and is not an instance of it. Is this not to attribute the property of expressing some property and lacking it? My rules seemed not to help with existential quantification. It was widely held that saying some F is a G is not to predicate being G of any F.

My answer was based on the relevance of the logical fact (IV): that asserting that some F is a G is the same as asserting, of everything whatever, that if everything other than it is not an F which is a G, then that thing is an F which is a G. This rule can be defended as following from a theorem of Lower Predicate Calculus (LPC):

$$((\exists x)(Fx \& Gx) \leftrightarrow (x)((y)(x \neq y \rightarrow \sim(Fy \& Gy)) \rightarrow (Fx \& Gx))).$$

But I do not want to be dependent on LPC and simply offer rule (IV). The rule shows how the earlier rules entail that calling a predicate t heterological merely denies that t is F, if F is the property expressed by t. If t expresses no property, then t is heterological for that reason.

For example, to say (truly) (HD): " 'dog' is heterological", is to say that some property is (1) expressed by "dog" and (2) not possessed by "dog". By (IV) that is to say of everything that if no other thing is a property expressed and lacked by "dog", then that thing is a property expressed and lacked by "dog". Now there is just one property expressed by "dog". (This is strictly speaking false---a log dog is not a canine. Ambiguity can be dealt with, but we ignore it for simplicity.) The conditional property attributed to everything by asserting HD makes no significant commitment, except for the case of the property of being a dog, which satisfies the antecedent. Our rules then entail that asserting HD is saying, of being a dog, that it is a property expressed by, but lacked by, "dog" which is to say that the word "dog" is not a dog, whether or not the assertor knows he has done this. Similar reasoning applies to all predicates. The crucial upshot is that, given any predicate that does express one property, it is not possible to say of it simply that there is a property it expresses and lacks, without actually attributing to it the negation of that property. It is possible to do this, and is done, if the predicate does not express a property. But when t expresses a property F, calling *t* heterological is to say that *t* is not F.

This means that "heterological", though perfectly meaningful, does not itself express a property, and is, for that reason, heterological. "Heterological" is (*roughly*) like "here". "My reading glasses are here" attributes a property to the glasses. Asserted this morning, it attributed a different property. "Here" is heterological because it does not uniformly express a property. (This is not ambiguity.) The self referentiality in the paradox is not the trouble---it is mistaking a common predicate for a common property---a conflation endemic among nominalists.

That conflation is also responsible for Russell's Paradox. Every property determines a set. By far the best understanding of set formation is that a set is the extension of a property. But what about the predicate (H)= "...is a non-self-member" (or the LPC predicate $\sim(x\varepsilon x)$)? If it expresses a property, then its extension would both belong and not belong to itself. Russell himself answered that there is no such property, but he had to also ban the predicate---rule it out of a logically proper language. The truth is that it is a fine predicate which does not uniformly express a property. Applied to the set of dogs, it says that set is not a dog. Applied to the set of cats, that it is not a cat etc..

These views are deeply in conflict with a highly influential doctrine formulated by Richard Montague, to the effect that there is no important theoretical difference between natural languages and the artificial languages of logicians. For those artificial "languages" are founded on uniform interpretations. In English, the property attributed by applying H depends on the subject to which it is applied. That variation is not allowed in the formal systems Montague favors.

This is difficult to illustrate because, for example, the LPC predicate $\sim(x\varepsilon x)$ is so different from an English predicate. It is "applied" to names, rather than to predicates. In some particular interpretation of a first order set theory which allows "impure" sets (sets with members that are not sets), the set of dogs might be named '*a*'. Instantiating to get $\sim(a\ \varepsilon\ a)$ does not convey the thought that the set of dogs is not a dog. Ironically, in such an interpretation, $\sim(x\varepsilon x)$ will determine a set. In any interpretation, any wf with one free variable determines a set. In a standard set theory, that set will not be allowed to be a member of the domain of the interpretation. That is an odd way of expressing the idea that not every property determines a set.

Similar problems arise in discussing the status of an attempt to formulate "A is not true" in the symbolism of LPC. There can be a predicate 'T' and the wf '~Ta', and an interpretation in which that wf is a member of the domain and is given the name '*a*'. But there is no way to accommodate the idea that the wf '~Ta', as asserted by that wf itself, is not true, while *we* could use the same wf to make a true assertion about it. This suggestion just seems hopelessly nonsensical. If you think there is no essential difference here from English, then my views will appear equally nonsensical.

Formal logic is based on sentences (closed wfs) and predicates (open wfs). Logic has to consider propositions and properties. It may be replied that this is only possible by considering sentences and predicates. That is true, but sentences and predicates may say, or attribute, various things. *What* they say (propositions) or attribute (properties) are different. This is the essential difference between natural language and "formalized language". A formalized language can be defined so as not to allow ambiguity. The multiple "interpretations" of LPC are radically different from "interpretation" in English. There is an uneliminable possibility that clever and creative users will exploit the established meaning of natural language expressions to give to some previously airy nothing a local habitation and a name. The elimination of this possibility is the mark that distinguishes formalized language. A fixed piece of English can be treated as a formal system and rules may be given for using the expressions which are readily understandable by English speakers. Breaking those rules will be, by definition of the system, failing to speak within the system. But English is not so definable. It has a vague identity determined by success in communication among users.

In *Philosophy of Logic* Quine says "Philosophers who favor propositions have said that propositions are needed because truth only of propositions, not of sentences, is intelligible. An unsympathetic answer is that we can explain truth of sentences to the propositionalist in his own terms: sentences are true whose meanings are true propositions. Any failure of intelligibility here is already his own fault." (page 10) Quine goes on to call the proposition view "...shabby...an imaginary projection from sentences". If this is right, then my line of response to the paradoxes is certainly shabby. But Quine's characterization of the relation between propositions and sentences arises from presuming that there is no such thing as what is said in dialogue and shared by understanding participants, which may then be summarized by a sentence which can serve as a memory aid, but could not be adequate for just any intelligent user of the same language. If we wish to know what, if anything, someone has said with the words "Justice is the interest of the stronger", to reply that it is the proposition that justice is the interest of the stronger, is at best, inapt.

Propositions do not have form in the same way that sentences do. A proposition which is the material disjunction of two propositions P, Q is the proposition which is the negation of the material conjunction of the material negations of P, Q. So a proposition is never simply a disjunction, or negation, etc. The purposes of the formal logic which has flourished so greatly are best served by things like sentences, not propositions. Philosophical dialogue is in terms of propositions. It originated in speaking and depended on memory. It is possible in writing, but harder because the temporal sequencing and persistence conditions are so different. Computers naturally work with things like sentences. The computational theory of mind and materialist nominalism generally reinforce formal logical theory in its emphasis on sentences and distrust of the idea of a proposition.

Sentences and their grammar are, of course, of great importance to Logic of any sort. By working with systems which are mathematically definable by their grammar or even by additional formal "semantics", significant mathematical results can be achieved. (Ironically, some of the deepest of these, Gödel's, are a support for antiformalist Platonism.) One notion definable for a formal system is completeness---the idea that every wf with the semantic property of validity has the syntactic property of provability. It is highly doubtful that this completeness can be given a good sense in terms of propositions. This brings me back to my thoughts about paradoxes.

One paradox offered as novel is Stephen Yablo's Infinite Liar, involving an infinite series of signs along a path. Sign 1 reads "Every sign along The Path numbered 2 or greater expresses a falsehood". Each successive sign n has the same words except for reading "numbered $n+1$ or greater". Familiar assumptions about what is said by each sign, determining this merely by indirect quotation of the words, quickly leads to a contradiction. This paradox is essentially similar to the Epimenides and is properly answered the same way. Each sign n says that $n+1$ is false and that $n+2$ is false, etc. *ad infinitum*. But $n+1$ says that $n+2$ is false. So what n says is inconsistent – that $n+2$ is false and that it is false that $n+2$ is false (that comes from denying $n+1$). So all the signs on the path are inconsistent. That is somewhat different from the inconsistency of the one Cretan statement, but the rules needed are the same.

However, a relative of Yablo's Paradox in terms of existential quantification is not answered by those rules. Make each sign n read "Some sign along The Path numbered $n+1$ or greater, expresses a falsehood". We get a contradiction following the old formulas for what is said. But the rules stated so far do not answer. This is true even of my wonderful rule for existential (Some F is G) claims, because it depends on there being either no F or just one, while this paradox has infinitely many equally qualified. This led me to note that the same trouble arises for a relative of the old Policeman. A dozen prisoners are forced to each make exactly one deposition, and they all depose "Something deposed by one of us is false".

My response was this rule V: to assert that some F is a G is to assert that if anything is such that nothing other than it is any better candidate for being an F that is a G than it is, then it is an F that is a G. Rule V shows all the depositions and signs along the path make inconsistent assertions and are thus false with no paradoxical claim to truth.

This sketch of my views was in response to a question as to my "main contributions to the philosophy of logic". If I have made any, they are like the things sketched here. From this, some will quickly infer that I have not made any. Rule V might be cited as a paradigm of the *ad hoc*. This reflects an ideal of completeness best suited to formal systems. From that perspective paradoxes are challenges to the formulation of exceptionless rules about validity and adjusting the rules of a system in response to bad results being proven is a mark of being *ad hoc*. By contrast, addressing paradoxes as they are presented is merely confronting particular arguments and assessing them carefully. True, I claim to draw general rules as lessons. But they are not offered as forestalling all possible paradoxes. It is no solution to a paradox to formalize a "language" in which it cannot be formulated, though this may contribute to valuable formal generality.

There were also questions about "the most significant advances" and "the most important open problems". I have no expectation of solving a problem listed as open, or discovering a new entry for the guild's list. There is rewarding, and valuable, work to be done in Logic, which need not qualify for such honors---and this is not to imply that the honors are anything but highly appropriate. I have managed to extricate myself, and some students, from some confusions. Sometimes this feels like significant advances, and remaining puzzles are open problems for any who share them.

A problem rather "open" for me is Burali-Forti's Paradox. Cantor's Paradox about the universal set (U) is based on "Cantor's Theorem". That can name a firm theorem in first order set theory, which has no such set, or a fully general proposition. That proposition is false about U, for essentially the same reasons that answer Russell's Paradox. U does contain its power set, *pace* Cantor's Theorem. There is no such property as being an element of U not belonging to its correlated set in P(U), any more than there is being a non-self-membered set. That makes the proof of Cantor's Theorem inapplicable to U. The set of all ordinals (which I must accept if there is such a property as being an ordinal) is more troubling than U. If (1) every set of ordinals is an ordinal and (2) every ordinal has a successor, there arises a contradiction. (2) is unassailable. (1) is a theorem in first order set theory. Or rather, (1) is a natural way of describing the theorem. It is the real content that is my main concern now.

7

Mark Colyvan

Professor
University of Sydney, Australia

---

1. Why were you initially drawn to the philosophy of logic?

I started out in mathematics and became interested in the notion of proof in mathematics. Like many mathematicians I thought that what constituted a proof of a mathematical result was cut and dried, until I found out about the intuitionist challenge to classical mathematics. This challenge fascinated me. I took some logic classes, looking for answers but found only more questions. It turned out that there was much more to philosophy of logic than the debate over intuitionistic logic and (so-called) classical logic. Who would have thought? There's modal logic, multi-valued logics, supervaluational logics, relevant logic, paraconsistent logics, and others. How is one supposed to choose between all these options? I was deeply puzzled, but hooked. Although I went on to write a PhD dissertation in the philosophy of mathematics, I found myself regularly returning to philosophy of logic, both as part of my work on philosophy of mathematics and for other reasons.

Another reason I was drawn to philosophy of logic was because of the logic community in Australasia. In Australasia non-classical logic is routinely taught and defended and debated enthusiastically. There is no presupposition that classical logic is the front runner or that philosophy of logic is about finding a way of reconciling classical logic with the various paradoxes. Philosophy of logic is taught as a living discipline with many important open questions. This community is very welcoming of new ideas and of new individuals. I learned so much from so many people: most notably Graham Priest, Ed Mares, Bob Meyer, Rod Girle, Greg Restall, Jc Beall, Norman Foo, John Slaney, Drew Khlentzos and others. With such teachers, colleagues, and, later, collaborators it was almost impossible not to be drawn into logic and philosophy of logic. Even as an undergraduate I had a real sense that this was an area where exciting things were happening and there were cool people making them happen.

2. What are your main contributions to the philosophy of logic?

Most of my work in philosophy of logic has, one way or another, been concerned with vagueness. I've written on the definition of 'vagueness' (Bueno and Colyvan 2012), on paraconsistent approaches to the sorites (Beall and Colyvan 2001; Hyde and Colyvan 2008), on similiarities between the sorites and the liar paradoxes (Colyvan 2009), and on generalizing the sorites paradox (Weber and Colyvan 2010). Let me say a little more about the last of these topics.

The sorites paradox is often presented in terms of a countable series of (totally-ordered) nearby states: 1 grain of sand, 2 grains of sand, 3 grains of sand and so on. We then note that the property "is a heap of sand" applies to at least one of these states and that if it applies to one state, it also applies to nearby states. The latter is crucial and is known as *Tolerance* (Wright 1975): small changes in state do not result in a change in the applicability of a vague property (or the corresponding vague predicate). It is interesting to note that the smaller the steps between the states, the more compelling the argument, yet in the limit, when the steps are infinitely small, as in continuous spaces, we were forced to discretise the space first (and hence present a less compelling version of the sorites argument). Consider a sorites series in terms of a woman's height and the property "tall". Clearly, a series of states where the steps between them are 0.5 meters is not compelling. We need the steps to be close: 1 mm, say. But 0.1 mm steps would be better still. Best of all would be to invoke the continuity of the underlying space and present a sorites without discrete steps at all (Chase, unpublished).

We can take things further still. We can abstract away from such metric spaces and formulate a sorites paradox in generalized topological spaces, where we only require the (topological) notion of connectedness. We can still formulate the crucial principle of Tolerance, and we do not need a metric to do this. This topological generalization of the sorites is useful in capturing family resemblance paradoxes (Weber and Colyvan 2010). For example, take a notion like "is a religion". There are various ways in which an activity can be more or less religion-like. The activity in question might involve, for example, ritualistic behaviour, a commitment to beings whose powers outstrip the average human's, the wearing of cerimonial attire, the singing or chanting of traditional pieces, and so forth. Most, if not all, of these components of religious activity come in degrees, though there is no obvious metric here and no obvious sequence of well-ordered steps from religion to non-religion. It is precisely for this reason that the traditional form of the sorites is not able to capture such family resemblance paradoxes. This is one place where the topological version of the sorites can be useful.

3. What is the proper role of philosophy of logic in relation to other disciplines, and to other branches of philosophy?

On the one hand, logic is the autonomous study of formal logical systems and philosophy of logic concerns itself with (amongst other things) debates about competing formal systems for specific tasks. On the other hand, logic can be thought of as a normative theory of deductive reasoning. On this latter conception, philosophy of logic is part of the much loftier enterprise of deciding how we should reason. Thus understood, philosophy of logic is central to all philosophy, indeed, it is central to all thinking. But whatever your view about the appropriate scope of logic, there seems little doubt that logic and philosophy of logic are important to many other disciplines. Here I'll give a couple of examples from my forays into applied philosophical logic.

I have an ongoing interest in the relationship between philosophy of logic and other disciplines such as mathematics and various branches of empirical science. For example, I've done work on problems arising from vagueness in conservation biology (Regan et al. 2000; Regan et al. 2002; Regan and Colyvan 2000), and how such considerations might give rise to the need for non-classical probability theory (Colyvan 2004; 2008). For example, a great deal of science is forced to trade in vague predicates. In conservation biology we have important categories such as "endangered", "critically endangered", and the like. In order to treat similar cases similarly, we need to respect the tolerance in the categories in question.

We need to classify species according to their risk of extinction and we use categories such as 'vulnerable', 'endangered', 'critically endangered' and the like. Despite the fact that these categories are vague, they are typically made precise by definitions involving rates of decline, population numbers and the like. But these definitions utilise cut-offs that are clearly arbitrary, and this can lead to problems. For instance, let's suppose that if the number of individuals in a population is less than n, then the species is to be classified as endangered. Let's further suppose that there is conservation funding available for breeding programs and the like for endangered species, but not for non-endangered species. Now consider some species with n individuals. According to the rules as they stand, the species is not endangered (according to the precisely defined use of 'endangered'), but presumably the species is not in good shape—it's almost endangered. What should we do? It might be in the species' best interest to kill a couple of them so that their numbers fall below n, then conservation funding for breeding programs can be put in place. But this is crazy. Surely it would be better to recognise the vagueness in the category 'endangered' and treat this vagueness directly using some of the tools logicians have developed for just such purposes: fuzzy logic, supervaluations, three-valued logics, and the like. With such tools at our disposal we no longer have to insist on sharp cut-offs in the categories in question; we can permit borderline cases of endangered and perhaps argue that such cases deserve some conservation funding, but less than the fully-endangered species. In short, we can treat like cases alike and respect tolerance in a way that is missing when we impose arbitrary cut-offs.

Another way vagueness gives rise to problems in conservation management and policy is related to probability theory. Standard (Kolmogorov) probability theory has classical logic written into its foundations: it is part of the tautologies of classical logic that are assigned maximal probability and if some proposition is assigned probability p, then the negation of the proposition in question is assigned 1-p. The latter is the probabilistic equivalent of the law of excluded middle. When we have vagueness in the mix, however, we have good reason to be suspicious of such classical assumptions. We might, for example, take some species to have a probability p of being endangered but deny that the probability, q, of it being not endangered is 1-p. We might, for example, take $p + q < 1$, with some of the probability density reserved for the borderline region of neither endangered nor not-endangered. Such considerations push for a non-classical probability theory (e.g. see Field 2000; Colyvan 2004, 2008; Shafer 1976; Walley 1991).

4. What have been the most significant advances in the philosophy of logic?

There have been so many; it is hard to know where to start. There is the development of modal logic in concert with philosophical applications, beginning with medieval logicians such as Scotus, Ockham, and Buridan, through C.I. Lewis to Prior, Barcan Marcus, Kripke, and others. There's the formalisation of logic by Boole, De Morgan, Frege, Peirce and others, the development of multi-valued logics and their applications by Łukasiewicz, Post, Kleene, Jaśkowski, and others. Without question, one of the most important advances was Gödel's incompleteness theorems and their philosophical consequences.

Many of these advances have come from thinking about the various logical paradoxes. So much of philosophy of logic, ever since Greek times, has been driven by paradoxes such as the liar paradox and the sorites. (The first known articulations of the sorites paradox together with an early presentation of the liar paradox are found in the work of 4th century BCE Megarian logician Eubulides of Miletus (Hyde 2011; Kneale and Kneale 1962).) A great deal of important work in logic and philosophy of logic has been directed at solving these paradoxes and great deal of important work has been inspired by the patterns of reasoning found in the paradoxical arguments themselves. Think, for example, of the way that liar-like reasoning features in Russell's paradox. Cantor employed similar reasoning to show that the cardinality of the power set of a set is strictly greater than the cardinality of the set (Cantor's theorem). And Gödel used liar-like reasoning in proving his famous incompleteness results. So a good case can be made that Eubulides made the most significant advances in philosophy of logic, with his early presentation of the liar paradox and, perhaps, the first presentation of the sorites paradox.

5. What are the most important open problems in philosophy of logic, and what are the prospects for progress?

There are too many to do justice to here. Solving the sorites paradox and the liar family of paradoxes are probably the big two. Even making significant inroads on these would be an outstanding achievement. It's not that there has been little or no progress in the last couple of thousand years on these open problems. In many ways it is the reverse problem: there have been so many solutions advanced and it is hard to know how to decide between them. It's not exactly a situation in which we have an embarrassment of riches. Each solution comes with some costs. I think it is fair to say that there are many candidates but no clear frontrunner.

I suspect that others in this volume will have much more to say about these two big open problems, so let me say a little about a less well-known open problem that is closely related: the problem of deciding when two paradoxes are to be treated by similar means. For example, in Eubulides of Miletus' list of puzzles we find The Bald Man puzzle and The Heap puzzle. The former is about hairs on a man's head and where to draw the line between bald and not-bald. The latter is about grains of sand and where to draw the line between heap and non-heap. These are clearly the same puzzle, despite the difference of subject matter and despite being given separate entries in the Eubulides list. Indeed, we usually treat all such puzzles as instances of the sorites paradox. What this means is that when someone proposes a solution to The Bald Man, the same solution must also apply to The Heap. We could even codify this bit of commonsense methodology into a principle: treat paradoxes of the same kind in the same way. This is usually called *The Principle of Uniform Solution* (Priest 1994).

I take it that the principle itself is uncontroversial enough, but there are devils in the details. First, it is worth noting that it is not a principle motivated merely by simplicity considerations. Rather, I take it that the idea is that one wants to treat the root of the problem. If a solution to

The Heap revolved around, for example, details of the metaphysical status of sand (or whatever) but failed to provide a solution to The Bald Man, it would be no solution at all. Without the Principle of Uniform Solution to keep us honest, we could easily slide into treating similar paradoxes differently and thus miss the core problem: whatever it is that is driving the paradox in question. To invoke a medical analogy, we would be treating the symptoms and not the underlying disease (Colyvan 2009). Still, the principle of uniform solution is not completely uncontroversial. I contend that everyone accepts this principle, although there is still room for serious disagreement over particular applications.

The example of The Heap and The Bald Man are clear cases where we have two paradoxes of the same kind. But not all such cases are so clear cut. For example, are Russell's paradox and the Liar paradox of the same kind? What of Curry's paradox and the liar? Yablo's paradox and the liar? What we would like are systematic ways to answer questions about when two paradoxes are of the same kind. There has been some progress on this topic, mostly focusing on the so-called paradoxes of self reference with the inclosure schema (Priest 1994; Smith 2000). But a general account of sameness of paradox is an outstanding open question. Moreover, it is an open question that will potentially have a significant impact on the way we approach the liar and the sorites. For example, if, as some have argued, the sorites and the liar paradoxes are more closely related than initially appears (Tappenden 1993; Field 2003; Priest 2010; Colyvan 2009), that would be a game changer. We would not only be obliged to find a solution to each paradox, but we would be obliged to provide *the same solution* to each.

The prospects for progress here, I think, are rather good. We already have a nice way of classifying the so-called of self reference with the inclosure schema (Priest 1994). I don't mean to suggest that this is the final answer here, but at the very least it is a good starting point and I think clearly captures something important about the structure of this class of paradoxes. There are significant questions about the appropriate level of abstraction: too high a level of abstraction and all the paradoxes look the same; too low a level of abstraction and there's too much irrelevant detail and all the paradoxes look different. There is also the issue of the best way to represent the structure of the sorites paradox and others. Again I think the prospects here are good. We have some proposals on the table (see Hyde 2011) and if nothing else, these can serve as useful points of departure.

Bibliography

Beall, Jc. and Colyvan, M. 2001. 'Heaps of Gluts and Hyde-ing the Sorites', *Mind*, 110(438): 401–408.

Bueno, O. and Colyvan, M. 2012. 'Just What is Vagueness?', *Ratio*, 25(1): 19–33.

Bueno, O. and Colyvan, M. 2004. 'Logical Non-Apriorism and the Law of Non-Contradiction', in G. Priest, JC Beall, and B. Armour-Garb (eds.), *The Law of Non-Contradiction: New Philosophical Essays*, Oxford University Press, 2004, pp. 156–175.

Chase, J. Unpublished. 'A Continuous Sorites', manuscript.

Colyvan, M. 2009. 'Vagueness and Truth', in H. Dyke (ed.), *From Truth to Reality: New Essays in Logic and Metaphysics*, Routledge, 2009, pp. 29–40.

Colyvan, M. 2004. 'The Philosophical Significance of Cox's Theorem', *International Journal of Approximate Reasoning*, 37(1): 71–85.

Colyvan, M 2008. 'Is Probability the Only Coherent Approach to Uncertainty?', *Risk Analysis*, 28(3): 645–652.

Colyvan, M. 2013. 'Idealisations in Normative Models', *Synthese*, 190(8): 1337–1350.

Field, H. 2000. 'Indeterminacy, Degree of Belief and Excluded Middle', *Noûs*, 34: 1–30.

Field, H. 2003. 'Semantic Paradoxes and the Paradoxes of Vagueness', in JC Beall and M. Glanzberg (eds.), *Liars and Heaps*, Oxford: Oxford University Press, Oxford, pp. 262–311.

Hyde, D. 1997. 'From Heaps and Gaps to Heaps of Gluts', *Mind*, 106: 641–660.

Hyde, D. 2011. 'Sorites Paradox', in E.N. Zalta (ed.), *The Stanford Encyclopedia of Philosophy* (Winter 2011 Edition), URL = <http://plato.stanford.edu/archives/win2011/entries/sorites-paradox/>.

Hyde, D. and Colyvan, M. 2008. 'Paraconsistent Vagueness: Why Not?', *The Australasian Journal of Logic*, 6: 107–121.

Kneale, W. and Kneale, M. 1962. *The Development of Logic*, Oxford: Oxford University Press.

Priest, G. 1994. 'The Structure of the Paradoxes of Self Reference', *Mind*, 103: 25–34.

Priest, G. 2010. 'Inclosures, Vagueness, and Self-Reference', *Notre Dame Journal of Formal Logic*, 51: 69–84.

Regan, H.M. and Colyvan, M. 2000. 'Fuzzy Sets and Threatened Species Classification', *Conservation Biology*, 14(4): 1197–1199.

Regan, H.M., Colyvan, M., and Burgman, M.A. 2000. 'A Proposal for

Fuzzy IUCN Categories and Criteria', *Biological Conservation*, 92(1): 101–108.

Regan, H.M., Colyvan, M. and Burgman, M.A. 2002. 'A Taxonomy and Treatment of Uncertainty for Ecology and Conservation Biology', *Ecological Applications*, 12(2): 618–628.

Shafer, G. 1976. *A Mathematical Theory of Evidence*, Princeton NJ: Princeton University Press

Smith, N.J.J. 2000. 'The Principle of Uniform Solution (of the Paradoxes of Self-Reference)', *Mind*, 109: 117–122.

Tappenden, J. 1993. 'The Liar and Sorites Paradoxes: Toward a Unified Treatment', *Journal of Philosophy*, 90: 551–577.

Walley, P. 1991. *Statistical Reasoning with Imprecise Probabilities*, London: Chapman and Hall.

Weber, Z. and Colyvan, M. 2010. 'A Topological Sorites', *The Journal of Philosophy*, 107(6): 311–325.

Wright, C., 1975. 'On the Coherence of Vague Predicates', *Synthese*, 30: 325–365.

8

Newton Carneiro Affonso da Costa

Professor of Philosophy
University of São Paulo, SP, Brazil

> As a lightning clears the air of unpalatable vapors, so an incisive paradox frees the human intelligence from the lethargic influence of latent and unsuspected assumptions. Paradox is the slayer of Prejudice.
>
> J. J. Sylvester

I summarize, in the remarks below, some of my views on the philosophy of logic. However, taking into account the limits of space that I have at my disposal, my remarks are schematic; details may be found in the references.

1. Why were you initially drawn to the philosophy of logic?

From the age of fifteen or sixteen, I have been interested in the problem of knowledge, its meaning, scope and limits. I perceived that I really needed to understand the problem of scientific knowledge. To do that, I dedicated myself to the study of certain areas of the scientific domain, in order to get a first-hand acquaintance with the foundations of logic, mathematics and some particular field of the empirical sciences. Among those sciences, I chose to concentrate on physics. Clearly, logic, deductive as well inductive, plays an essential role in the theoretical basis of all sciences, including mathematics.

A good amount of general philosophy and, in particular, of epistemology, was also required. Amongst other things, I studied various contemporary philosophers, like Russell, Carnap, Quine, Enriques and Popper. The influence of French thinkers on my philosophical development was also very important. Deserving of particular mention would be Descartes, Brunschvicg, Cavaillès and Lautman.

Mathematics, philosophy and physics showed me the relevance of non-classical logics. For example, intuitionistic logic is indispensable to the comprehension of constructive mathematics, and quantum mechanics presents logical difficulties that have suggested the creation of quantum logics. Even though one may defend the thesis that heterodox logics are epistemologically irrelevant, one has to discuss the status of most new logics. Both the philosophy of science and more general philosophical problems almost impose on us the view that some new logics are true logics, such as tense logic and paraconsistent logic are true logics. But it is clear that other non-classical logics, like non-structural logics, nonlinear logic and the logic of default, are only tools associated with questions of information processing, i.e., to computing, at least today.

We may say, exaggerating a little bit, that the present situation, in the field of logic, forces us to recognize that a revolution is occurring in that field analogous to the revolution caused by the birth of non-Euclidean geometries in the 19th century.

Moreover, certain other aspects of the history of logic in the last century also helped to push me to the study of logic and its philosophy. For example, logic, which was, since the time of Aristotle till the beginning of the 20th century, a source of mathematically trivial results, became, little by little, a very complex mathematical edifice. Topics like forcing, forking, Gödel's theorems, algebraic logic, and others showed that logic could be mathematically significant, giving rise to fundamental inquiries and results as profound as the most important mathematical ones. These developments originated new conceptually significant problems in the spheres of logic, its applications and philosophy.

To confirm my observations on the progress in the domain of logic, it is sufficient to quote from Yu. I. Manin's excellent book *A Course in Mathematical Logic for Mathematicians*, Springer, 2010, page VII:

> In the intervening three decades [since 1977], a lot of interesting things happened to mathematical logic: (i) Model theory has shown that insights acquired in the study of formal languages could be used fruitfully in solving problems of conventional mathematics. (ii) Mathematics has been and is moving with growing acceleration from the set-theoretic language of structures to the language and intuition of (higher) categories, leaving behind old concerns about infinities: a new view of foundations is now emerging. (iii) Computer science, a no-nonsense child of the abstract computability theory, has been creatively dealing with old challenges and providing new ones, such as the P/NP problem.

Independently of the total correctness of Manin's assertions, they demonstrate, at least in outline, that profound changes have occurred in logic in the last decades.

2. What are your main contributions to the philosophy of logic?

To begin with, I insist on the fact that my conceptions related to the philosophy of logic are inseparable from my logic work. Therefore, some discussion of the latter is essential to explain the former.

Leaving aside some incursions into the domains of induction and tense logic, deontic and modal logics, the logics of belief and of justification, and algebraic logic, my main areas of research in logic are the following: 1) Paraconsistent and non-reflexive logics; 2) Quasi-truth; 3) Generalized Galois theory and its applications to the foundation of physics; 4) The theory of valuations; 5) Recursion theory and complexity.

There is in [7] a good description of the nature of paraconsistent logic; its authors write that, on pages 791 and 792:

> In a few words, paraconsistent logics (PL) are the logics of inconsistent but nontrivial theories. A deductive theory is paraconsistent if its underlying logic is paraconsistent. A theory is inconsistent if there is a formula (a grammatically well-formed expression of its language) such that the formula and its negation are both theorems of the theory; otherwise, the theory is called consistent. A theory is trivial if all formulas of its language are theorems. Roughly speaking, in a trivial theory 'everything' (expressed in its language) can be proved. If the underlying logic of a theory is classical logic, or even any of the standard logical systems like intuitionistic logic, inconsistency entails triviality, and conversely. So, how can we speak of inconsistent but nontrivial theories? Of course, by changing the underlying logic to one which admits inconsistency without making the system trivial. Paraconsistent logics do just this job.
>
> Our use of terms like 'consistency', 'inconsistency', 'contradictory' and similar ones is syntactical, which is in accordance with the original metamathematical terminology of Hilbert and his school. In order to treat such terms from a semantic point of view, in the field of paraconsistency, one must be

able to build, first, a paraconsistent set theory. This is possible, as we will see, although most semantics for paraconsistent logics are classical, i.e., constructed inside classical set theories. So, to begin with, it is best to employ the above terms syntactically.

Inconsistencies appear in various levels of discussion of science and philosophy. For instance, Peirce's world of 'signs' (which we inhabit) is an inconsistent and incomplete world. Bohr's theory of the atom is one of the well-known examples in science of an inconsistent theory. The old quantum theory of black-body radiation, Newtonian cosmology, the (early) theory of infinitesimals in the calculus, the Dirac δ-function, Stokes analysis of pendulum motion, Michelson's 'single-ray' analysis of the Michelson-Morley interferometer arrangement, among others, can also be considered as cases of inconsistencies in science. Given cases such as these, it seems clear that we should not eliminate a priori inconsistent theories, but rather investigate them. In this context, paraconsistent logics acquire a fundamental role within science itself as well as in its philosophy. As we will see below, due to the wide range of applications which nowadays have been found for these logics, they have an important role in applied science as well.

Paraconsistent logic helped make it clear that the domain of rationality has a wider scope than the traditional one: logic may cope with contradiction without annihilation of the discourse. Although negation inside a paraconsistent logical system cannot be classical, nonetheless it possesses most formal properties of classical negation, and thus deserves to be considered a different kind of negation. Strong systems of paraconsistent logic exist and even a paraconsistent mathematics is being developed. In addition, there are various applications of paraconsistent logic, inclusive to real situations, such as those of robotics, distribution of energy, and traffic control in large cities. Today, if one is willing to systematize all natural sciences in a single axiomatic system, this requires a paraconsistent logic, since, for example, quantum mechanics and general relativity are incompatible. (By the way, in some applications the paraconsistent systems are employed as mere technical classical tools to organize inference.) Other references on paraconsistent logic are the following: [1], [2], [3] and [5].

Non-reflexive logic was born to cope with two basic kinds of questions: those of causal inference and those raised by quantum mechanics. I quote the following passage connected with this type of logic:

> It is a striking feature of research in logic in the twentieth century that a plurality of logics has emerged. A similar pattern can be found in the case of several logics. What was initially taken to be an unchallenged principle of classical logic turned out to be open to revision in light of the introduction of a suitable logic. For example, excluded middle, an important principle in classical logic, does not hold in general in intuitionistic logic or in many-valued logic. The principle of non-contradiction, another central principle of classical logic, need not hold in paraconsistent logic... but this still leaves open the principle of identity. This principle is challenged by non-reflexive logics.
>
> Roughly speaking, non-reflexive logics are logics in which the principle of identity does not hold in general. One of the main motivations for the construction of these logics (and it turns out that are infinitely many of them) emerges from the foundations of physics. According to certain interpretations of quantum mechanics (such as the one favored by Schrödinger) it does not make sense to attribute identity to quantum particles. If this is indeed the case... the principle of identity seems to fail. ([4], p. 182)

Some authors call the law that any proposition implies itself the principle of propositional identity. One of the categories of non-reflexive logic includes logics with implications that don't satisfy this principle, also called the reflexive law of implication. For example, causal implication and explanatory implications are non-reflexive.

Quasi-truth and generalized Galois theory (or general theory of set-theoretic structures) constitute the basis for a structural systematization of science, especially physics. Quasi-truth generalizes Tarki's conception of truth, and its logic is paraconsistent, which should be expected, since theories such as general relativity and quantum mechanics are incompatible. So, the logic of quasi-truth may be seen as the starting point for an axiomatic reconstruction of present-day empirical science (see [6], [8], [10] and [13]).

The theory of valuations deals with the subject of a general semantics for logical systems. It includes the standard Tarskian semantics, but appeals to the syntactic level of the language employed. All logics are complete and correct in relation to their semantics of valuations. In various cases, for instance that of Brouwer-Heyting propositional logic, the method of valuations provides us with new decision methods (see [2] and [7]).

My work on recursion theory and complexity allows us to determine some new limits of the axiomatic method in connection with the foundations of the natural sciences, as well as of mathematics (some references are [11] and [14]).

The sciences aspire to provide objective knowledge. It seems that logic (and its philosophy) entails that scientific knowledge is a kind of belief that is justifiably quasi-true, i.e., objective in certain sense.

Paraconsistent and non-reflexive logics are the starting point of some of my views related to the philosophy of logic.

As I noted, quantum theory and general relativity are logically incompatible. In addition, all physical theories are approximate, and even if true according to the correspondence theory of truth, we would be unable to know that it is so. Therefore, in the foundations of the empirical sciences, we need a new version of truth that I called quasi-truth. It, among other things, generalizes Tarski's definition of truth and consists of a sort of 'as if' truth. The most interesting aspect of it is that the logic of quasi-truth is paraconsistent (see [6] and [8]).

Non-reflexive logic makes evident that logic has connections with experience, i.e., it isn't entirely a priori. Quantum theory, for instance, drives us to the necessity of a profound analysis of some heterodox logics. I believe that one of the characteristics of contemporary logic is the edification and discussion of new logics, in particular of heterodox logics.

Paraconsistent and non-reflexive logics motivated a profound analysis of logical concepts such as those of negation, conjunction, quantification, and property, including sets and classes. In particular, various groups of negation are possible and convenient.

To summarize, my principal contributions to the philosophy of logic are the following:

1. The analysis of the meaning of non-classical logics, in particular for science, with emphasis on physics.

2. The investigation of negation and other logical concepts taking into account their roles in non-classical logics.

3. The introduction of the concept of quasi-truth and its logic as a new basis for scientific theories, which can be applied even in mathematics, especially in the domain of so-called experimental mathematics.

4. A distinctive approach to inductive logic (which, among other things includes statistics) based, in part, on subjective probability (details in [8] and [9]). My work on recursion theory and complexity was important in making explicit some limitations essentially involving the scientific method (compare with [11]).

5. The philosophical elucidation of various semantic notions by means of the theory of valuations, above all the notions of correctness and of completeness of a logical system.

3. What is the proper role of philosophy of logic in relation to other disciplines, and to other branches of philosophy?

Philosophy of logic is mainly the critical study of the foundations of logic and of its signification for human knowledge in general. In consequence, one of its principal tasks is to contribute to a better understanding of mathematical and scientific knowledge. In this sense, logic helps the philosopher in his critical analysis of different domains of knowledge and the systematization of those domains via the axiomatic method.

Let us take a look at two examples: a) In quantum mechanics there exist various central problems, including those of the nature, meaning, and scope of the relation of identity in connection with quantum objects. Non-reflexive and non-distributive logics were born to help explain some difficulties involving the identity of quantum particles. b) Some forms of metaphysics, for instance structural metaphysics, are related to the logical and mathematical theory of set-theoretic structures (see [11] and [10]). In philosophy of logic we are concerned with those and similar subjects.

The relevance of logic and its philosophy is very well understood with reference to pure mathematics. In applied mathematics, say in information processing and practical computing, which gave birth to several new logics, I had to submit those logics to a detailed philosophical investigation. The significance of such themes becomes clear when we see that they have links to some of the most influential disciplines of our time such as robotics, Artificial Intelligence and neuroscience.

We live in the era of computing, information theory and robotics. Progress in some present-day fields of knowledge is exponential, and, to have a deep understanding of what is occurring we need, surely, to appeal to logic and its philosophy.

4. What have been the most significant advances in the philosophy of logic?

In the history of logic numerous significant advances could be here listed and discussed, such as those associated with the names of Aristotle, Boole, Frege and Russell. Restricting myself to the last one hundred years, I think that the main advances in philosophy are the following:

1. The philosophical developments motivated by the works of Gödel and Tarski.
2. The philosophical results originating from non-classical logics, above all the heterodox logics.
3. The inquiries into the logical foundation of quantum mechanics.
4. The definition of quasi-truth and the investigation of its logic.

A few comments on such advances are in order.

Gödel's investigations in logic and in the foundations of mathematics culminated in his incompleteness theorems, which brought about a revolution in logic and in the foundations of mathematics. His philosophical treatment of the notions of finitary method and of the constructivity of mathematical procedures, his inquiries in the foundations of set theory, and his philosophical reflexions on the meaning of the incompleteness theorems deserve mention. On this last topic, Gödel says that,

> It is *this* theorem [the second incompleteness theorem] which makes the incompletability of mathematics particularly evident. For, *it makes it impossible that someone should set up a certain well-defined system of axioms and rules and consistently make the following assertion about it: All of these axioms and rules I perceive (with mathematical certitude) to be correct, and moreover I believe that they contain all of mathematics.* If somebody makes such statement he contradicts himself. For if he perceives the axioms under consideration to be correct, he also perceives (with the same certainty) that they are consistent. Hence he has a mathematical insight not derivable from the axioms. (K. Gödel, Collected Works, vol. III, p. 309.)

The notion of truth is one of the central ideas underlying mathematics and the sciences, as well as logic. In formulating a rigorous concept of truth, Tarski mathematized a notion which caused problems in the philosophy of mathematics and of the sciences in general. Tarski's account of truth fits like a glove the intuitive, informal concept of classical truth, constituting a remarkable advance in both logic and epistemology. Moreover, it was the starting point of numerous extensions of the concept of truth, extensions that, in some cases, are more adapted to the treatment of various questions related to the natural sciences, like the idea of quasi-truth.

I already made reference to the significance of the birth of non-classical logics, especially of alternative logics. Thus, intuitionistic logic and its variant versions contributed to the clarification of the meaning of constructivity in mathematics and of information processes linked, say, to Turing machines. Paraconsistent logic showed the existence of diverse forms of weak negation and the possibility of a new treatment of contradiction. Some systems of tense logic yield a deeper philosophical analysis of the concept of time. Non-reflexive logic is going to give rise to novel ideas in the domain of philosophy (details in [4]).

5. What are the most important open problems in philosophy of logic, and what are the prospects for progress?

I shall answer this question by limiting myself to my current research. One of the principal open problems I am studying is related to the genuine meaning of non-classical logics, in particular of heterodox logics. Have such logics an epistemological significance, are they authentic logics, or are they just mathematical tools founded on classical logic? I believe that the future progress of knowledge, above all of quantum mechanics, will clear up these questions. However, we need to learn much more of quantum mechanics, quantum logic and related matters, to surmount these difficulties.

A second subject that must be philosophically explained is the relation between logic and space-time. The common systems of logic are built independently of considerations of space and time. But how is this legitimate? If logic is structured as if neither space nor time exists, then how can it be applied to objects that are in space and time? In fact, we have various space-times in theoretical physics: for example, Newtonian space-time (used in classical mechanics and non-relativistic quantum mechanics), Minkowski space-time (used in special relativity and quantum field theory) and the space-times of general relativity (used in quantum field theory on curved spaces and cosmological models). A philosophical examination of the precise links between space-time and logic is essential.

A third interesting problem is the following: I conceive logic as a discipline incorporating what is usually called inductive logic, in which statistics is included. Today, a serious philosophical difficulty with respect to this part of logic consists in the following: the presupposed concept of probability, needed by non-relativistic quantum mechanics, is not the common one of standard probability theory and statistics. In effect, in quantum mechanics there isn't any Kolmogorov basic structure, since the set of events don't satisfy the conditions required by the customary theory of probability. In addition, the standard rule of conditionalization doesn't work. In order to understand quantum mechanics, as well as most of its interpretations, do we need a new theory of probabilities and statistics? The philosophical investigations of these topics are progressing and more progress is expected.

I have just listed the three preceding problems mainly because I think that logic is not a subject entirely justifiable by itself. On the contrary, its value is, in a great part, a consequence of its connections with the world of mathematics and of science (in particular of physics) and also with everyday inferences. So, logic's main philosophical relevance depends on issues involving, in part, other regions of knowledge.

Anyhow, I would like to emphasize that progress in philosophy means above all a better understanding of the problems examined. As Bertrand Russell would say, the goal of philosophy is to replace inarticulate beliefs by articulate disbeliefs.

Bibliography

[1] da Costa, N. C. A., 'On the theory of inconsistent formal systems', *Notre Dame Journal of Formal Logic* XV (4), 497-510, 1974.

[2] da Costa, N. C. A., Logiques Classiques et non Classiques. Masson, Paris, 1997.

[3] da Costa, N. C. A. and V. S. Subrahmanian, 'Paraconsistent logics as a formalism for reasoning about inconsistent knowledge bases', Artificial Intelligence in Medicine, 1, 167-174, 1989.

[4] da Costa, N. C. A. and O. Bueno, 'Non-reflexive logics', Revista Brasileira de Filosofia, 58 (232), 181-196, 2009.

[5] da Costa, N. C. A. and N. Grana, Il Recupero dell'inconsistenza. L'Orientale Editrice, Napoli, 2009.

[6] da Costa, N. C. A., O. Bueno and S. French, 'The logic of pragmatic truth', *Journal of Philosophical Logic*. 27, 603-620, 1998.

[7] da Costa, N. C. A., D. Krause and O. Bueno, 'Paraconsistent logics and paraconsistency'. Handbook of the Philosophy of Science. Philosophy of Logic, Dale Jacquette editor, Elsevier, 2007, pp. 791-911.

[8] da Costa, N. C. A. and S. French, 'Pragmatic truth and the logic of induction', *British Journal for the Philosophy of Science*, 40, 333-356, 1989.

[9] da Costa, N. C. A. and S. French, 'On Russell's principle of induction', Synthese 86, 285-295, 1991.

[10] da Costa, N. C. A. and S. French, Science and Partial Truth. Oxford University Press, 2003.

[11] da Costa, N. C. A. and F. A. Doria, On the Foundations of Science. Coppe, Rio de Janeiro, 2008.

[12] French, S. and D. Krause, Identity in Physics: A Historical, Philosophical and Formal Analysis. Oxford University Press, 2010.

[13] Mikenberg, I., N. C. A. da Costa and R. Chuaqui, 'Pragmatic truth and approximation of truth', Journal of Symbolic Logic, 1986, 201-221.

[14] Chaitin, G., N. da Costa and F. A. Doria, Gödel's Way. Taylor and Francis, London, 2012.

9

Pascal Engel

Directeur d'études, École des hautes études en sciences sociales, Paris and honorary professor, Université de Geneve

---

CONFESSIONS OF A CLASSICAL NORMATIVIST

1. Why were you initially drawn to the philosophy of logic?
My interests have always been those of a philosopher, who took logic both as a tool and a source of problems for philosophy, not those of a practicing logician. I once tried, under the guidance of Jean Van Heijenoort, to work on the French logician Jacques Herbrand, whose pioneering work laid the basis of proof-theory, but I soon had to realize my limitations. I got interested in logic because issues such as nominalism *vs.* realism about universals, the nature of truth and of propositions seemed to me more salient and tractable when raised in the context of logical theories. Indeed I have never seen any real opposition between philosophy and logic, broadly conceived as the set of issues dealing with the most abstract parts of our thinking and of our language. I drew my initial inspiration from Dummett's monumental commentary on Frege, which seemed to me to be a kind of *Critique of Logical Reason*, and got interested in Davidson's program in semantics and in the opposition between realist and antirealist theories of meaning. Dummett strongly emphasized the connection between intuitionistic logic and antirealism, argued for molecularism on the basis of his conception of logical constants, and defended a verificationist conception of meaning. I sided with Davidson, advocating a realist conception of semantics and classical logic, but I have always admired the way Dummett approached and shaped these problems. Like everyone else, I took Quine and Tarski to be the gospel, but I have always been attracted by the style of English logicians and philosophers - Prior, Geach, Dummett and Strawson – in part because they had a better sense of the history and a wider metaphysical scope. Indeed, Quine too had an interest in ontology, but he wanted a minimal ontology. Early on I was interested in the work of Ruth Marcus, who was not only a great logician, but also had great philosophical ideas (about modalities, reference and belief) which were very much underappreciated at the time. I am still nostalgic

about the 1970s, which were a Golden Age for analytic philosophy. At that time logic and the philosophy of logic were not very distinct from the philosophy of language, and philosophers, linguists, and logicians talked a lot to each other and often coexisted harmoniously in philosophy departments. Today, most practicing logicians are in computer science departments, and the philosophy of language has lost its empire. Indeed those who today call themselves "formal philosophers", and whose mother tongues are advanced logic, computer programming and probability theory, are often dismissive of "straight" analytic philosophers, who are suspected of sloppiness, because they do not always express their views within a formalism (see for instance Clark Glymour's "manifesto" and Timothy Williamson's "Must do better"[1]). In the 1970s, I heard people like Patrick Suppes and Dana Scott express similar feelings about analytic philosophy, but the divide between the formalists and the informalists was not so great at that time.

2. What are your main contributions to the philosophy of logic?

It's not for me to say. But if I were to characterize my approach I would say that what is distinctive of it is that it tries to deal with the central issues of the philosophy of logic in a synthetic way. In that respect, I am not a very analytic philosopher. Although logic has been the main tool and source of inspiration of analytic philosophers during the past century, it seems to me that attention to detail, to puzzles and to (mostly logical) paradoxes has tended to obscure the wider issues which are at stake: How general is logic? How much is formal? Does it tell us anything about reality? What is its relationship to thought and to thinking? To language and meaning? Is there anything like logical knowledge and in what does it consist? Which logic is the right logic? In my book *The Norm of Truth: An Introduction to the Philosophy of Logic* (Prentice Hall 1991), I considered the basic problems of logic as centered around three main questions, each associated with a specific paradox or problem:

- How can logic be informative (Mill's paradox: if the premises of a syllogism already contain the information present in the conclusion how can we learn from logical inferences)? How can there be logical *knowledge*?

- How can the laws of logic be justified in a non-circular way (Agrippa's trilemma for logical knowledge: either the justification is circular, or it is arbitrary, or it leads to an infinite regress)?

---

[1] Timothy Williamson, "Must do Better", in *The Philosophy of Philosophy*, Blackwell, Oxford 2005, Clark Glymour , http://choiceandinference.com/2011/12/23/in-light-of-some-recent-discussion-over-at-new-apps-i-bring-you-clark-glymours-manifesto/

- How can logic be normative (Lewis Carroll's paradox of inference: how can logical laws or rules move us?)

I have been interested in these three issues ever since. Concerning (a), it is often tempting to answer the problem of informativeness by saying that triviality and topic neutrality are mostly features of elementary and first-order logic: as soon as logic goes beyond the first-order, with modal, higher-order and non-classical logics, inference becomes a much less straightforward matter, and logical structures become complex and "interesting", in contrast to the dull monotony (in both senses of this word) of classical logic. This is indeed the same complaint raised by mathematicians: the poorer the logic, the more boring logical inferences are. I disagree. Classical first-order logic can be interesting, and its structures can be complex and not trivial. Gentzen's calculi are not trivial, and exhibit the shape of proofs in a beautiful way, Herbrand's theorem too is not trivial. Wittgenstein said that there cannot be surprises in logic. That is just false. There is such a thing as logical knowledge and we can learn by deduction. And it is not true that in order to gain insight we need to adopt some non-classical logic. One can learn from classical logic and a classical approach. Very often in science a simpler theory has more explanatory payoffs when it deals with complex issues than a more complex theory, which posits more entities and more sophisticated explanations. Thus attempts made philosophers like Davidson or linguists like James Higginbotham to analyze our event language in classical quantificational terms, or Williamson's epistemicist's theory of vagueness, which stick to classical logic, seem to me more interesting than theories which at once posit more complicated structures, such as higher-order quantification and supervaluations respectively. We learn more when a classical scheme cannot be applied to, for instance, natural language than when many non-classical schemes apply. Of course, the love of classical simplicity has its diminishing returns, and there is a point where one has to go non-classical in logic. But classical logic is the norm. However interesting and creative the efforts of dialetheists, paraconsistent and dynamic logicians can be, it seems to me that classical logic remains, and has to remain, the standard tool. That may seem very conservative, given the blooming of non-classical logics. But I am an absolutist in logic: although non-classical logics are very interesting, it seems to me that only classical logic can serve as a norm for philosophical inquiry. This may not be true for purposes other than philosophical – especially in mathematics, economic modeling and computer science. But when it comes to philosophy, we have to stick to bivalence. Contradictions cannot be true. Hegel will never triumph over Russell.

With respect to (b), philosophy of logic is the mirror of general epistemology. Just as we need a theory of the justification of our basic beliefs, in particular those based on perception, we need an account of what Crispin Wright has called basic logical knowledge. And the options here are very close to those of Agrippa's trilemma: logical laws or rules are primitive and based on nothing else, or they are circularly based on other laws and rules, or there is an infinite regress. If one rejects these options, there is no choice but to accept that logical laws are based on nothing. One may thus adopt skepticism or conventionalism. Although a lot of thinkers have been tempted, in one way or another, by the last option, including Carnap, whose principle of tolerance says that "in logic there are no morals and everyone is free to choose his own system", I think that we have to resist this extreme relativism. I hate the kind of sloppiness which pervades all present day logic with respect to which system is best. Everyone seems to assume that we are in a kind of supermarket where you are free to choose whichever logic suits our particular purposes. But that is not true. In logic there are morals, and we are not free to choose. One system has to be the best and we need foundations. So I reject the kind of logical pluralism which seems to be accepted by most practicing logicians.

Now, if one believes, as I do, that there is but one logic which is the right logic, which one is it? Is it intuitionistic logic? I agree with Dummett that we need a justification of deduction, and that we cannot rest content with a form of holism where logical rules support each other by a kind of network association. Some inference rules have to be primitive and basic. Dummett and Prawitz have argued in favor of a form of logical foundationalism about the logical constants, and claimed that we need to impose certain conditions, such as harmony and conservativeness on logical connectives in order to avoid weird logical connectives like Prior's infamous *tonk*. But according to them, these constraints imply logical revisionism and the choice of intuitionistic logic as the right logic. But do the tighter constraints on logical connectives imply logical revisionism? It is far from clear. Alan Weir (1986), Christopher Peacocke (1987, 1993) and Peter Milne (1994) have argued that harmony is available to the classicist too. So one can be a logical realist, a partisan of bivalence and of classical logic, and also adopt the canons for logical constanthood emphasized by the intuitionist. Am I, then, a classicist of the strongest stripe, like Timothy Williamson? Yes I am. But being a classicist need not entail agreement with all the claims of the ultra-conservative view that Williamson advocates. He famously says that "When philosophical considerations lead someone to propose a revision of basic logic, the philosophy is more likely to be at fault than the logic". I do not see why philosophy ought to be ruled by logic in such a way. Williamson probably means that logic wears the trousers because philosophy by itself, lacking logical rigor, is unable to state what is right in terms of justification of logical rules. I disagree. There can be *philosophical* arguments for conservatism. One is Quine's "meaning variance thesis", according to which the non-classical logician "changes the subject" by giving new meanings to the logical connectives. But this view depends strongly on Quine's views on radical translation, which one needn't accept (and I don't). Another kind of argument is suggested by Williamson when he says about non-classical treatments of vagueness: "Conditional proof, argument by cases and *reductio ad absurdum* [these are all invalidated by supervaluationism] play a vital role in systems of natural deduction, the formal systems closest to our informal deductions. [...] Thus supervaluationists invalidate our natural mode of deductive thinking". Now to what extent is a mode of thinking "natural"? Cognitive psychology suggests that it is far from clear that humans follow the rules of ordinary logic, and there are many studies which seem to show that inferences like modus ponens, modus ponens, or disjunctive syllogism are often violated. So we cannot be content with an argument to the effect that our "natural" ways of thinking favor classicism. A better argument is, according to me, that classical logic is normative for our thinking. A norm is an idealization. An idealization is not a natural law, be it psychological or physical. But it's not independent of the facts. One line to take here is to adopt a form of reflective equilibrium conception of logical laws, similar to the one taken by Rawls in ethics. I find congenial the comparison between logic and ethics, which has been with us since Herbart ("logic is the ethics of thought"). But neither moral principles nor logical principles are a matter of revision and of consensus. They have be firm and, to speak like Frege, as solid as a rock.

Question (c) is the one which has occupied me most. I have been interested in the nature of normativity in logic, but also in epistemology and in the philosophy of mind, and indeed in ethics, and my work in all these fields is a reflection about the nature of norms. There are a number of analogies between all these domains, and the structure of the normative domain is strongly unified, but it also displays important dissimilarities. In a series of essays and in a forthcoming book in French[2], I have examined these issues, as they arise, in my view from

---

[2] "Logical Reasons", *Philosophical Explorations*, 8, 1, march 2005, 21-35, "Dummett, Achilles and the Tortoise", *The Philosophy of Michael Dummett*, The Library of Living Philosophers 2007; "Oh! Carroll! Raisons, normes et inférence " in *Klèsis*, 13 , 2009 ; "How to resist a Tortoise", "Wie man einer Schildkröte widersteht", *Proc. Deutscher Kongress 2011*, Meiner 2012, « The lessons of Carroll's regress », 2012, to appear in *the Carrollian, Avatars de la tortue*, to appear.

Lewis Carroll's famous and enigmatic dialogue between Achilles and the Tortoise, published in *Mind* in 1895. Why does the Tortoise refuse to draw the conclusion of a simple inference in *modus ponens* form? Is it, as the usual lesson of the tale has it, that he conflates a premise and a rule of inference? Or is it that he refuses to take logical laws (or rules) as normative and capable of moving our minds? The Tortoise's problem is logical *akrasia*: he sees the rule, but willing refuses to follow it. He doubts that logic is normative, or doubts that logical norms have a motivational power. To answer his challenge, one has to give an account of the normativity of logic. The norm cannot be a further proposition that we consciously and reflectively entertain – for otherwise we would be led to Carroll's regress. It cannot be that one has to follow the rule blindly without thinking about it, for in logical reasoning, we attend to reasons. Neither is it that we have a form of *knowledge-how* related to the inference form, or that we master a practice, because logical knowledge is not a species of know how. None of these solutions work. I try to defend a more complex picture. Our logical knowledge is based on a tacit knowledge of rules (largely unconscious), but we also develop rational dispositions associated with our main logical concepts; in inferring we attend, although non-reflectively, to logical reasons. We have to combine such an account with a realist view of logical reasons and of epistemic norms.

3. What is the proper role of philosophy of logic in relation to other disciplines, and to other branches of philosophy?

The philosophy of logic is neither a branch of logic (it's not "philosophical logic", understood as a logical inquiry about matters more or less philosophical) nor a branch of the philosophy of language, nor a branch of the philosophy of science, although it shares with these fields of inquiry a number of themes and concerns. The philosophy of logic deals with the philosophical problems raised by logic. A number of these problems are epistemological. What is the justification of logical inferences? What kind of knowledge is logical knowledge? What is a proof? Other problems are ontological. What is logic about? To what kind of entities is the logician committed? To what extent can there be a formal ontology? Other problems are closer to the philosophy of language. What is predication? What is logical form? Others are closer to the philosophy of mathematics and computer science. So in my view, the philosophy of logic is a part of philosophy, not of logic. It cannot be itself "formal" in the way a logical theory can be formalized, although it has to use, to a large extent, formal results and logical theories.

A number of philosophers do not see things that way. They practice what they call "formal philosophy", "formal epistemology", "deontic logic" or "formal value theory", and advocate the use of formal models throughout philosophy. They hold that one cannot deal with a philosophical problem or concept without first translating it into formal terms and then constructing a logical theory, from which one can derive various theorems. This approach has proved extremely fruitful in a number of domains, *e.g.* for the notion of truth (Tarski), for the formalization of the ontological proof for God's existence (Gödel, Plantinga, Oppy), for giving a model of the origins of the social contract (Skyrms) and for modeling belief change (Alchouron-Gardenförs-Mackinson, Levi, Rott). This is to name only a few successful formal approaches to specific philosophical problems. These methods are descendants of the axiomatic method although the formalisms that they use are much more complex than those of logic proper, for they often use modal logic, model theory, probability theory, dynamic logic, higher-order logics and a wealth of other formalisms. The formalisms provide a great deal of clarification: one sees exactly what the assumptions are, and what follows from them. One shows that certain *prima facie* attractive ideas are problematic (for instance Lewis's impossibility results about conditionals), and so one is able to test, through logic, the solidity of certain philosophical proposals. Although I admire these methods, and support strongly their use, I am skeptical about what they can really achieve.[3] Very often I find the basis of the formal language used, for instance for modeling belief change, very stipulative, and so I have doubts about what they actually prove. For instance Hans Rott, in his monumental work on belief change,[4] suggests that his results show that theoretical reason is largely a part of practical reason. For all the impressive results that he demonstrates, it seems to me that the claim is premature. Tarski's formalization of the ordinary notion of truth is a major contribution to logic and to philosophy, but it hardly solves the main issues about truth, which continue to be as debated as they were when he first proposed his semantic conception in the 1930. I also find fascinating recent work on Fitch's paradox of knowability, which purports to show that on an antirealist view of truth, all truths are known. But it is still an open question whether all truths are knowable or not. So, when Timothy Williamson urges his colleagues to use more the rigorous methods of logic to deal with philosophical problems, I approve the advice, but I am pessimistic about the possibility that formal methods can actually solve, or even

---

[3] See P. Engel "Formal methods in philosophy: Shooting right without collateral damage", in n T. Czarnecki, K. Kijania- Placek, O. Poller, & J. Woleński (Eds.), *The Analytic Way: Proceedings of the 6th European Congress of Analytic Philosophy*, 2010.

[4] H. Rott, *Change, Choice and Inference: A Study of Belief Revision and Nonmonotonic Reasoning*, Oxford University Press, Oxford, 2001.

make philosophical problems more tractable. Logical modeling cannot replace philosophy. I am not, however, a partisan of what used to be called "informal logic", in particular in the hands of P.F. Strawson, who took it as a kind of investigation into the structures of natural language, and as opposed to formal logic. I believe, on the contrary, in the power of formalism, but I also believe that there are limits to its use and to its capacity to solve philosophical problems.

4. What have been the most significant advances in the philosophy of logic?

During the last part of the twentieth century, the work of Michael Dummett and Dag Prawitz on truth, proof and logical consequence stands as the most significant for the philosophy of logic, together with the discussions of sequent calculi and structural logics, which give us a much deeper view about the structure of inference and about the nature of logical constants especially by Stephen Read, Neil Tennant and Stewart Shapiro. I am impressed by work on substructural logics, relevant logics, and linear logics, although, as I said above, I resist the kind of logical pluralism which is defended today by writers like Graham Priest, Greg Restall and Jc Beall.

I have always been an admirer of Crispin Wright's work in the philosophy of logic and mathematics, his renovated Frege-program, although I do not share his commitments. I also admire a lot Hintikka's system of ideas, which goes with a vast program of logical reform. But I must admit that I have never been tempted to work within it, except perhaps when it comes to the reading of the history of philosophy, with issues about time, modality and necessity.

One of the most important developments in the philosophy of logic during the last fifty years has been the renewal of the great tradition of logical metaphysics, which goes from Leibniz and Bolzano to Husserl and early analytic philosophy. This tradition is in permanent opposition to the empiricist tradition of Hume, Mill and Quine. Although the Quine's influence has been overwhelming, it's fair to say that Quine's doubts and qualms about modal logic have been overcome by the work of Ruth Marcus, David Lewis and Saul Kripke. Today quantified modal logic is a basic tool of philosophers, and the issues at the intersection of logic and metaphysics are at the center of the field, especially in the work of Kit Fine. But modal logic is not in opposition to classical logic. It is an extension of classical logic than a revision of it. As the title of Williamson's last book has it, we can conceive of modal logic as metaphysics by other means.

The other developments which I find very significant are those which relate to epistemology. A large part of the philosophy of logic is epistemology. Issues about the nature of belief, of knowledge, about conditionals and the nature of reasoning, as well as issues about epistemic normativity are as central in the epistemology of logic as in general epistemology. Work in epistemic logic and in formal epistemology is here very relevant, and Timothy Williamson's "knowledge first" program is very important, not only because of Williamson's classical absolutist stance, which I share, but also because it promises to illuminate the notion of logical knowledge. On these matters, I also find very important Christopher Peacocke's work on reason and the *a priori*, and Paul Boghossian's work on the nature of inference. Ian Rumfit is also a philosopher of logic whose work I admire a lot. The work of Igor Douven on probability, assertion and conditionals seems to me also first rate, as does Erik Olsson's work in formal epistemology, and David Christensen's work on rationality and belief.

In the philosophy of mathematics, the work of Paul Benacerraf, George Boolos, Hartry Field, and Penelope Maddy, among others has been very important. Among present day philosophers of mathematics I consider the work of Leon Horsten, Volker Halbach, Jeffrey Ketland on the relationship between truth and proof to be very impressive.

5. What are the most important open problems in philosophy of logic, and what are the prospects for progress?

Is logic formal? Which logic is the right logic? What is the nature of consequence? What is an inference? What justifies logical laws and rules? What is logical knowledge? What is existence and to what extent does quantification tell us what it is? What are propositions? What is predication? What is truth? What is logical truth? Is logic normative? In spite of a wealth of formal developments, none of those basic problems seem to me solved, and there is no hope of solving them, because they are, like all philosophical problems, wide open. It is interesting to see how old issues resurface. For instance, the present debates concerning deflationism about truth echo Carnap's neutralism about ontology, and the debates about the meaning of logical constants renew the issue of the analytic/synthetic distinction. This does not mean that we cannot have progress, for the developments in logic help us see these issues in novel ways, and by introducing new methods and concepts.

Besides the work mentioned above, one issue which I still find very important is the relationship between logic and psychology. Frege and Husserl successfully fought against the psychologism of their time, but psychologism is still alive. Pluralism about logic favors a descriptive stance about natural language semantics, and many philosophers adopt psychological theories of reasoning and of meaning based on impressive advances in the psychology of reasoning. We are still far from having satisfactory evolutionary conceptions of the origins of logic. The idea that logic originated in dialectics and in argumentation, rather than in abstract thought about deduction and truth, is still very attractive to a number of logicians and cognitive scientists, who reject what they call normativism. I am a normativist, in the old fashioned style. I take logic to be normative, absolute, classical. But that does not mean that we should be content with its pure theory without trying to understand how logic is normative and how logic regulates our thinking. To understand logical normativity, we have to attend to psychology, without forgetting that it's logic, and not psychology, that wears the trousers.

11

Susan Haack

Distinguished Professor in the Humanities, Cooper Senior Scholar in Arts and Sciences, Professor of Philosophy and Professor of Law, University of Miami, USA.

---

*FIVE ANSWERS ON PHILOSOPHY OF LOGIC**

> *We come to the full possession of our power of drawing inferences, the last of all our faculties; for it is not so much a natural gift as a long and difficult art.*
>
> —C. S. Peirce[1]

1. Why were you initially drawn to philosophy of logic?

In my student days—though it was by then (up to a point, and still somewhat grudgingly) admitted that women maybe *could* do philosophy—the usual assumption was that we were more suited to the supposedly "softer" side of the discipline, ethics in particular. But I found ethics formidably difficult; indeed, I still fondly recall, as a B.Phil. student in Oxford, writing a paper on deontic logic for a tutorial on moral philosophy with Philippa Foot—and her kind response: "yes, I see, *this* is more your kind of thing."

So, at that time, perhaps I was drawn to philosophy of logic in part out of a temperamental resistance to those thoughtless assumptions about women's supposed intellectual bent; in part because questions in this area seemed exactly hard enough to be genuinely challenging, but so not slippery and evanescent as to elude my grasp entirely; and in part, of course, because as I began to read Frege's, Russell's, Tarski's, Quine's and, a little later, Peirce's writings on the subject, I found so much to think about.

---

* © 2013 Susan Haack. All rights reserved.
[1] C. S. Peirce, *Collected Papers*, eds. Charles Hartshorne, Paul Weiss, and (vols. 7 and 8), Arthur Burks (Cambridge, MA: Harvard University Press, 1931-58), 5.358-59 (1877).

2. What are your main contributions to philosophy of logic?

I'll start with the Ph.D. dissertation that became my first book, *Deviant Logic*.[2] This is, as the saying goes, a young man's book. But there's a lot in it: an examination of the distinction between deviant logics (systems with the same vocabulary as classical logic, but different theorems and/or valid inferences) and extended logics (systems with additional vocabulary and additional theorems and/or valid inferences involving that new vocabulary);[3] a diagnosis of what goes wrong with Quine's confused, and confusing, arguments that deviant logicians (a.k.a. "pre-logical peoples") are "a myth invented by bad translators";[4] an exploration of the understandings (and misunderstandings) of truth behind various deviant systems; and chapters on future contingents, Intuitionism, vagueness, reference failure, and even quantum mechanics.

This first book remains in print, now in an expanded edition with a longer title, *Deviant Logic, Fuzzy Logic: Beyond the Formalism*.[5] The longer title tells a story: a reviewer of the original edition had pointed out that, though the book covered a lot of ground, it didn't include fuzzy logic. In fact, I'd never even *heard* of fuzzy logic; so I went straight to the library to check it out. This was in the early days of computer searches; and I still remember how I chuckled when I read the opening line of the first article on the reading list a librarian compiled for me: "In this paper we will discuss modal logic and probability theory, *but we will not discuss fuzzy logic*." Other things on the list proved more informative, however; and in due course I would write critical papers both on fuzzy logic and on the idea that truth is a matter of degree.

Fuzzy logic is described by its inventor, electrical engineer Lotfi Zadeh, as a logic in which truth-values are fuzzy, local, and subjective, the set of truth-values is not closed under the usual propositional operations, and "linguistic approximations" have to be introduced to guarantee closure; in which inference is approximate rather than exact, and semantic rather than syntactic; and completeness, consistency, axiomatization, and proof-procedures are "peripheral."[6] But this sacrifices all the virtues that Frege wanted formal logic *for*. Moreover, when you

---

[2] *Deviant Logic* (Cambridge: Cambridge University Press, 1974).

[3] *Id.*, chapter 1.

[4] W. V. Quine, *Word and Object* (New York: Wiley, 1960), p.387. Haack, *Deviant Logic* (note 2 above), 8-10. See also Haack, "Analyticity and Logical Truth in *The Roots of Reference*" (1977), reprinted in Haack, *Deviant Logic, Fuzzy Logic: Beyond the Formalism* (Chicago: University of Chicago Press, 1996), 214-31.

[5] Haack, *Deviant Logic, Fuzzy Logic* (note 4 above).

[6] Lotfi Zadeh, "Fuzzy Logic and Approximate Reasoning," *Synthese* 30 (1975): 407-25.

read the fine print you realize that the real work is being done by informal linguistic analysis, and the elaborate formal apparatus is largely redundant; and you notice that that, despite his insistence that fuzzy logic is itself vague, Zadeh ends up imposing a completely artificial precision: truth is defined as: "0.3/0.6 + 0.5/0.7 + 0.9/0.9+ 1/1"—i.e., as the fuzzy set to which degree of truth 0.6 belongs to degree 0.3, degree of truth 0.7 to degree 0.5, ..., etc.; and "very true" is defined as "true squared"(!).[7] In any case, Zadeh's underlying idea, that "true" is vague, is the result of his first misconstruing legitimate locutions like "very true," "quite true," and then compounding the mistake by introducing such bizarre locutions as "rather true" and "fairly true."[8]

Some defenders of fuzzy logic objected that I just *had to be* mistaken; after all, they argued, in its electrical-engineering applications, fuzzy logic *works*. So in the new edition of *Deviant Logic* I added an explanation of the workings of "fuzzy controllers" for air-conditioning systems and the like, showing that they *don't*, in fact, rely on fuzzy logic.[9] So when, shortly after this second edition appeared, I received a mysterious package from Bart Kosko, I held it to my ear to make sure it wasn't ticking—but no, it wasn't a bomb, but a copy of his enthusiastic book about fuzzy logic, inscribed "to Susan Haack, with warm fuzzy feelings." (So far as I know, Prof. Zadeh has never responded to my critique of fuzzy logic, nor to my comments on fuzzy controllers; but to this day he will occasionally send me little puff-pieces about the wonders of fuzzy engineering.)

As a graduate student in Oxford, I taught elementary logic; as a college lecturer in Cambridge, I made a deal with Renford Bambrough: I would teach the young men from St. John's logic, if he would teach the young ladies from New Hall ethics;[10] and then for many years I taught a year-long course on philosophy of logic at the University of Warwick. Before long, though, I began to chafe at the lack of a suitable textbook; which was how I came to write my second book, *Philosophy of Logics*.[11]

---

[7] Susan Haack, "Do We Need 'Fuzzy Logic?'" (1979), reprinted in *Deviant Logic, Fuzzy Logic* (note 4 above), 232-42.

[8] *Id.*, 240-42; Susan Haack, "Is Truth Flat or Bumpy?" (1980), reprinted in *Deviant Logic, Fuzzy Logic* (note 4 above), 243-58.

[9] Haack, *Deviant Logic, Fuzzy Logic* (note 4 above), pp. 230-31. (Perhaps needless to say, I have no background in electrical engineering; so it took me most of a very long, and very hot, summer to figure this out!)

[10] Among those "young men from St. John's" was Graham Priest, whom I taught logic, my first year in Cambridge, from the propositional calculus through Gödel's theorem. I am not, however, responsible for the dialethic logic for which he is now known, which I presume was due to the influence of Richard Routley.

[11] Susan Haack, *Philosophy of Logics* (Cambridge: Cambridge University Press, 1978).

This book has also proved long-lived; and, in its Spanish, Italian, Portuguese, Korean, and Chinese editions,[12] has been used around the world.[13] Almost everywhere I give lectures, it seems, someone in the audience was brought up on "Phyllis" (as this book is affectionately known at home). Especially memorable was a 2008 visit to the University of Valparaíso, Chile, where the philosophy department had used the book (the faculty the English edition, the students the Spanish edition) since its publication, and where I gave a lecture entitled "*Filosofía de las Lógicas*, Trente Años Después," explaining how I would write the book today; and a recent e-mail from a forensic scientist in England—in response to my request to an electronic list of people in the area for information about how fingerprint-matching software works—an e-mail that asked: "Are you THE Susan Haack, the one who wrote *Philosophy of Logics*?" Ermm, well, yes.

Like the new title of the second edition of *Deviant Logic*, the plural "logics" in the title of *Philosophy of Logics* tells a story. In the concluding chapter,[14] I carefully disentangle the central metaphysical and epistemological questions about logic, and give my answers. On the metaphysical side, I articulate a tentative defense of a kind of global pluralism. And on the epistemological side, in a paper from the same period, "The Justification of Deduction,"[15] I argue that problems analogous to those that notoriously arise in the attempt to justify induction also arise in attempts to justify deduction.

There's a lot else in Phyllis, too: including chapters on the distinctions between logic, philosophy of logic, and meta-logic, on validity, sentence connectives, quantifiers, singular terms, truth-bearers, theories of truth, paradoxes, modal logics, and many-valued logics. Maybe it's worth mentioning specifically my diagnosis of Quine's objections to modal logic,[16] and my exposition of Tarski's theory of truth.[17] Patiently working through the tangle of issues about Tarski's theory and its philo-

---

[12] There were, of course, the usual pitfalls of philosophical translation: the Spanish edition, for example, translated "relevance logicians" as "lógicos relevantes," making them seem more important than I believe them to be; the Portuguese translation had F. P. Ramsey using an analogy, not from cricket, but from baseball; and the Italian edition made the memorable mistake of translating "rat," in Quine's observation that "rat" is not semantically part of "'rat,'" any more than it is of "Socrates."

[13] And as I write this, a projected French translation has, at last, just begun.

[14] *Philosophy of Logics* (note 11 above), chapter 12.

[15] Susan Haack, "The Justification of Deduction" (1976), reprinted in *Deviant Logic, Fuzzy Logic* (note 4 above), 183-91. This paper has by now been reprinted several more times, and in 2013 appeared in Spanish.

[16] Haack, *Philosophy of Logics* (note 11 above), pp.178-187.

[17] *Id.*, pp.99-127.

sophical implications was hard work; but, to be candid, at the time I didn't think it especially remarkable. Now, however, when Tarski is routinely described as a correspondence theorist, or as a deflationist, or a disquotationalist, or as having given a theory of the truth of propositions, or, etc., my exposition seems like a more important contribution to keeping the record straight than I dreamt at the time.

Realizing that my earlier distinction of deviant vs. extended systems needed modification to acknowledge that some relevance logics, for example, are *both* deviant *and* extended, I was led to some serious thinking about the concept of relevance; which, I came to see, is not a formal but a material concept. This, I now believe, undermines the hope of a formal logic of relevance. It also helps explain how Kuhn arrived at the mistaken idea that standards of quality of evidence are paradigm-relative;[18] and it sheds light on the concept of relevance crucial to evidence law.[19]

As my philosophical interests have grown, I have turned my attention to other areas, writing books on epistemology[20] and philosophy of science,[21] and numerous articles in these fields and in philosophy of language, metaphysics, social philosophy, etc., and yes, even in ethics—in papers on the ethics of research (1996),[22] on affirmative action (1998),[23] and on academic virtues (2010).[24] And by now I've been drawn into other fields too—notably, the law. Some years ago I started still-ongoing work on issues involving evidence, proof, and scientific testimony, and on legal philosophy more generally. But I've not left my interest in philosophy of logic behind; I have written a whole series of

---

[18] Susan Haack, *Defending Science—Within Reason: Between Scientism and Cynicism* (Amherst, NY: Prometheus Books, 2003), pp.76-77.

[19] Susan Haack, "Legal Probabilism: An Epistemological Dissent" (first published, in Spanish, in 2013), in Susan Haack, *Evidence Matters: Science, Proof, and Truth in the Law* (New York: Cambridge University Press, 2014), pp. 47-61.

[20] Susan Haack, *Evidence and Inquiry* (Oxford: Blackwell, 1993; 2nd, expanded edition, Amherst, NY: Prometheus Books, 2009).

[21] Haack, *Defending Science—Within Reason* (note 18 above).

[22] Susan Haack, "Preposterism and Its Consequences" (1996), in *Manifesto of a Passionate Moderate: Unfashionable Essays* (Chicago: University of Chicago Press, 1998), 188-208.

[23] Susan Haack, "The Best Man for the Job May be a Woman ... and Other Alien Thoughts on Affirmative Action," in Haack, *Manifesto of a Passionate Moderate* (note 22 above), 167-87.

[24] Susan Haack, "Out of Step: Academic Ethics in a Preposterous Environment" (2010), in Haack, *Putting Philosophy to Work: Inquiry and Its Place in Culture* (Amherst, NY: Prometheus Books, 2008; expanded ed. 2013), 251-67 (text) and 313-17 (notes).

papers on truth, for example; a paper on Peirce and logicism;[25] a piece on formal methods in philosophy;[26] and a much-downloaded study of the place of logic (including deontic logic!) in the law.[27]

The "truth" series began with two pieces defending the legitimacy of the concept;[28] and continued with two more articulating the ramifications of the distinction between truth, i.e., the phenomenon, and truths, i.e., particular true claims, beliefs, propositions, etc. While there are many truths, I argued, there is only one truth;[29] while some truths are vague, truth is not a matter of degree; while some truths are made true by things people do, truth is objective; while some truths make sense only relativized to a place, time, or jurisdiction, truth is not relative; and while some propositions are only partly true, truth does not decompose into parts.[30] In 1974 I had shown that Post's non-standard many-valued logic serves to represent partial truth in the sense of "part of p is true";[31] in 2008 I also explored the other meaning of "p is partially true," "p is part of the truth.[32] Again, in 1974 I had written at length about (what I would now call) the logical conception of precision;[33] in 2008 I also explored another kind, the poetic.[34] The same year I published a paper comparing truth in science and in the law,[35] and by 2010 I was ready to present a full-dress account of legal truth.[36]

And my longstanding interest in the scope and limits of logic has lately begun to bear new fruit. In *Defending Science*, I showed that, and why, formal-logical models of scientific reasoning, whether inductivist,

---

[25] Susan Haack, "Peirce and Logicism: Notes towards an Exposition," *Transactions of the C. S. Peirce Society* 29.1 (1992): 33-56 (text) and 301-13 (notes).

[26] Susan Haack, "Formal Philosophy? A Plea for Pluralism" (2005), in *Putting Philosophy to Work* (note 24 above), 235-50 (text) and 301-313 (notes).

[27] Susan Haack, "On Logic in the Law: 'Something, but not All,'" *Ratio Juris* 20.1 (2007): 1-31.

[28] "Confessions of an Old-Fashioned Prig," in *Manifesto of a Passionate Moderate* (note 22 above), 7-30; "Staying for an Answer: The Untidy Process of Groping for Truth (1999), in *Putting Philosophy to Work* (note 24 above), 35-52.

[29] "The Unity of Truth and the Plurality of Truths" (2005), in *Putting Philosophy to Work* (note 24 above), 53-65 (text) and 271-73 (notes).

[30] Haack, "The Whole Truth and Nothing but the Truth," *Midwest Studies in Philosophy* XXXIII (2008): 20-35.

[31] Haack, *Deviant Logic* (note 4 above), 62-63.

[32] Haack, "The Whole Truth and Nothing but the Truth" (note 30 above), 28-29.

[33] Haack, *Deviant Logic* (note 4 above), chapter 6.

[34] Haack, "The Whole Truth and Nothing but the Truth" (note 30 above), 25-28.

[35] "Of Truth, in Science and in Law," *Brooklyn Law Review* 73.2 (2008): 985-1008.

[36] Susan Haack, "Nothing Fancy: Some Simple Truths about Truth in the Law" (2010) in Haack, *Evidence Matters* (note 19 above), 294-324.

deductivist, or probabilistic, must fail; for the "grue" paradox teaches us that such reasoning relies, not on form alone, but on the relation of scientific vocabularies to real kinds of thing and stuff in the world.[37] In "On Logic in the Law,"[38] I showed that, and why, formal-logical models are also inadequate to capture legal reasoning; for such reasoning inevitably involves stretching and adapting legal concepts as society, technology, manufacturing, etc., change. And in "The Growth of Meaning and the Limits of Formalism,"[39] with the help both of Peirce and of Oliver Wendell Holmes, I developed an approach to meaning that unified these two lines of argument.

3. What is the proper role of philosophy of logic in relation to other disciplines, and to other branches of philosophy?

As my answer will reveal, I'm uneasy with the implication of uniqueness in "*the proper role.*"

Let me begin by distinguishing two uses or senses of the word "logic": a broad, in which it refers to the theory of whatever is good in the way of reasoning ("LOGIC"), and a narrow, in which it is restricted to the syntactically characterizable aspects of good reasoning ("*logic*").[40] LOGIC, so conceived, includes both *logic* and philosophy of *logic*—as one sees in Peirce's writings. This broad conception can still be found in, e.g., Dewey's *Logic: The Theory of Inquiry*.[41] But in the wake of Frege's by now hugely influential work in *logic*, the narrow conception has become predominant.

As the title of Dewey's book suggests, LOGIC would include in its very broad scope at least much of what would today be thought of as epistemology, philosophy of science, etc. But the question of the relations of *logic* to other fields is very different, and far from straightforward.

As I explained in the last paragraph of the previous answer, *logic* falls well short of exhausting what can be said about the quality of reasoning either in the sciences, or in legal arguments—or, I will now add, in philosophy. Yes, occasionally a philosopher will make a formal-logical error. For example, as I argued in *Deviant Logic*, Aristotle's argument that future-contingent statements are neither true nor false rests on a modal fallacy;[42] and, as I argued in *Evidence and Inquiry*, Davidson's

---

[37] Haack, *Defending Science* (note 18 above), pp. 40, 52, 84-86.

[38] Haack, "On Logic in the Law" (note 27 above).

[39] "The Growth of Meaning and the Limits of Formalism," *Análisis Filosófico* XXIX.1 (May 2009): 5-29.

[40] As I did, for example, in "On Logic in the Law" (note 27 above), 9-10.

[41] John Dewey, *Logic: The Theory of Inquiry* (New York: Henry Holt, 1938).

[42] *Deviant Logic* (note 4 above), 77-78, 80-81.

Omniscient Interpreter argument that our beliefs are mostly true does, too.[43] (Aristotle argues from "Necessarily, if it is true that there will be a sea battle tomorrow, then there will be a sea battle tomorrow" to "If it is true that there will be a sea battle tomorrow, then necessarily there will be a sea battle tomorrow." Davidson argues from "It's impossible that there be an omniscient interpreter unless people's beliefs are mostly true," and "It's possible that there's an omniscient interpreter" to "People's beliefs are mostly true.") But far more often, in my experience anyway, problems in philosophical arguments are likely to be the result of unnoticed ambiguities,[44] flabby concepts, untenable dualisms, false presuppositions, and the like.

If logicism had been a viable account of mathematics, philosophy of mathematics would be a branch of philosophy of *logic*. But I don't believe logicism *is* viable. Similarly, if natural-kind terms were rigid designators, at least a significant part of philosophy of science would, again, be a branch of philosophy of *logic*. But I don't believe natural-kind terms *are* rigid designators; on the contrary, I believe that they have meanings, and that these meanings grow as our knowledge of the world grows.[45]

Nor am I convinced that formal-logical tools offer more than very limited help in our understanding of natural languages. The collapse of the "Davidson program" shows that Tarski was right all along to insist that rigorous formal methods like his apply only to well-behaved formal languages, and aren't suitable to natural languages like English or Polish.[46] Davidson himself would eventually conclude that there is no such thing as a language, in the sense that he and many philosophers of language had assumed.[47] My view is that what we loosely call a natural language is really better conceived as a kind of federation of similar-enough idiolects, and that how similar is "similar enough" depends on the task at hand.[48]

---

[43] Donald Davidson, "A Coherence Theory of Truth and Knowledge" (1983), in Alan Malachowski, ed., Reading Rorty (Oxford: Blackwell, 1990), 120-34, p.131. Haack, *Evidence and Inquiry* (note 20 above), pp.105-06.

[44] The subject of my B.Phil. dissertation, by the way, was ambiguity and its consequences in philosophy.

[45] "The Growth of Meaning and the Limits of Formalism" (note 39 above), §2.

[46] Alfred Tarski, "The Concept of Truth in Formalised Languages" (1933), in Tarski, *Logic, Semantics, Metamathematics*, trans. J. H. Woodger (Oxford: Clarendon Press, 1956), 152-78, p.165.

[47] Donald Davidson, "A Nice Derangement of Epitaphs," in Ernest Lepore, ed., *Truth and Interpretation* (Oxford: Blackwell, 1986), 433-46, pp. 445-46.

[48] Susan Haack, "The Growth of Meaning and the Limits of Formalism" (note 39 above), §1; "Belief in Naturalism: An Epistemologist's Philosophy of Mind,"

4. What have been the most significant advances in philosophy of logic?

I'm puzzled by this question: am I being asked to talk about the most significant advances in philosophy of logic *ever*, or just about what has happened recently?

Obviously, sketching even a few of the important advances made by Aristotle, Frege, or Peirce would take far more words than I have. And, while it's quite possible that somewhere in the world there's a new Peirce or a new Frege, working in obscurity as they did, whose thought is as ground-breaking as theirs was, if so, sadly, I'm not familiar with his or her work.

What I am familiar with, I'm afraid, is that philosophy of logic doesn't seem to be bucking recent trends in philosophy more generally. Just like other areas, it seems to be becoming more and more detached from its own history, more and more fragmented, more and more cliquish, more and more self-absorbed, and more and more inclined to set older problems aside unresolved as attention shifts to a new fad. And while I'm sure there's worthwhile work out there, the pressure to publish is now so severe, and the volume of publications so bloated, that it's nearly impossible to find the good stuff among the dross. That said, the footnotes in my next answer will mention some recent work I think is promising.

5. What are the most important open problems in philosophy of logic, and what are the prospects for progress?

I'm as uncomfortable with "*the most important*" as I was with "the proper role"; so I'll begin by saying that on such crucial matters as truth, meaning, modality, and the grounds of logic there's a lot more work to be done; and then go on to list some of the topics on which I'd like to have a better theoretical grasp than I believe we now have—a list in which, not surprisingly, Peirce will loom large.

- In *Philosophy of Logics* (like everyone else at that time), I called Ramsey's a "redundancy theory" of truth. By now, however, when all his papers on truth are available, and it's clear that Ramsey *didn't* think "true" was redundant, I prefer "laconicism."[49] More importantly, it's well worth exploring the questions that Ramsey's account left open, among them: whether there is an adequate understanding of propositional quantifiers (as in, "for some p, Plato said that p, and p") that doesn't itself rely on the concept of truth;

---

[49] F.P. Ramsey, *On Truth*, eds. Nicholas Rescher and Ulrich Majer (Dordrecht, the Netherlands: Kluwer, 1992). The word "laconicism" was coined by Kiriake Xerohemona.

and what a detailed laconicist approach to the semantic paradoxes would look like.[50]

- As the last point suggests, one consequence of the outstanding difficulties in the theory of truth is that we don't yet have a complete understanding either of the source of the semantic paradoxes, or of the most appropriate response. In this context I will mention Peirce's remarkable presentation and diagnosis of what we would now call the Strengthened Liar. Peirce begins with two columns of parallel arguments: one from the premise that "this proposition is not true" is *true*, to the conclusion that it is *not true*, and the other from the premise that "this proposition is not true" is *not true*, to the conclusion that it is *true*. Every step is valid, Peirce argues; so the source of the problem must be their only shared premise, that "whatever is said in the proposition is that it is not true." This, he concludes, is false; rather, like every proposition, the paradoxical proposition *also* asserts its own truth.[51]

- This analysis of the Strengthened Liar takes up just a few pages of Peirce's early paper "The Grounds of Validity of the Laws of Logic." [52] While at this time (1868) the logic Peirce uses is still syllogistic,[53] his approach to this issue is a good deal more sophisticated than, for example, the conventionalism that would come later. Further thought about Peirce's arguments specifically, and about the still-unresolved question of the grounds of logic more generally, would be welcome.

- While Peirce is on my mind, I'll also mention that, though modern modal logic has its roots in C. I. Lewis's work, Peirce had long before represented modal arguments in his "Gamma Graphs."[54]

---

[50] There seems to have been some progress on this, in the work of Arthur Prior, C. J. F. Williams, and María-José Frápolli. See Frápolli, "The Logical Enquiry into Truth," *History and Philosophy of Logic* 17 (1996): 179-97.

[51] Peirce, *Collected Papers* (note 1 above), 5.340-41 (1868).

[52] *Id.*, 5.318-57 (1868).

[53] In 1902—after he had made the logical innovations that led him, a few years later but independently of Frege, to a unified propositional and predicate calculus—Peirce raises another excellent question: "Why Study Logic?", and lists ten assumptions that must be true if such study is to be worthwhile: e.g., that there is objective truth, and that good reasoning can lead to it. *Id.*, 2.119-216 (1902).

[54] The "gamma graphs," as the name suggests, refers to the third part of his "existential graphs," a diagrammatic logical notation. See *Collected Papers* (note 1 above), 4.510-29 (1902), and 4.573-84 (1906). In 1903 Peirce writes that "possibility and necessity are relative to the state of information" (4.517); but by 1906 he acknowledges that "a mere possibility may be quite real" (4.580).

Perhaps because his diagrammatic notation, very intuitive at the propositional level, is pretty complex at the level of quantifiers, and formidable at the modal level, Peirce's approach to modal logic seems to have been little explored; but a serious study might prove as worthwhile as an examination of his 1909 experiment in three-valued logic turned out to be.[55]

- One consequence of the dynamic approach to language proposed in "The Growth of Meaning" was that what statements are analytic (or express analytic propositions, if you prefer) changes over time. Shocking? Not really. In Shakespeare's day, when "silly" meant "simple" and "sooth" —as in "soothsayer"— meant "truth," "silly sooth is simple truth," which is now pretty much meaningless, was analytic. But I have as yet only a very incomplete understanding of the consequences of this temporal relativity, or of what exactly a neo-descriptivist conception of naming might look like if we took the growth-of-meaning idea to heart.[56]

There's nothing canonical about this list; but I've already used more words than I was allotted, so I'll stop here.[57]

---

[55] Robert E. Lane, "Peirce's Triadic Logic Revisited," *Transactions of the C. S. Peirce Society*, 35 (2), 1999: 284-311.

[56] Here, I think, some progress has been made by Chen Bo in his "Kripke's Semantic Argument against Descriptivism," *Croatian Journal of Philosophy* XIII, no.39 (2013): 421-45, and two as-yet unpublished papers, "Social Constructivism of Language and Meaning" and "Socio-Historical Causal Descriptivism: An Alternative Theory of Names."

[57] My thanks to Mark Migotti for his helpful comments on a draft.

10
Jaakko J. Hintikka

Professor
Collegium for Advanced Studies, University of Helsinki

0.

The five questions one is supposed to answer here carry certain presuppositions that I do not fully share. The most basic one is that there is a distinguishable discipline called 'philosophical logic'. Of course, one can separate fields of research and give them names, but one source of trouble with philosophy of logic, thought of perhaps as the study of the conceptual foundations of logic, has been pursued without taking sufficiently into account other scholarly enterprises.

1. Why were you initially drawn to philosophy of logic?

Hence, I cannot say why I was initially "drawn" to the philosophy of logic. My work in and on logic led me naturally deeper and deeper into the conceptual foundations and conceptual problems. Sufficiently general and philosophically relevant questions of this kind might and also will here be called "philosophy of logic", but that should not imply any sharp borderline or even division of labor.

2. What are your main contributions to philosophy of logic?

Also, there seems to be an unspoken assumption to the effect that philosophical logic, whatever else it is or may be, is in a state of normal science, so that one can speak of its "advances", "contributions" to it, its "proper role" and its "open problems". If there is a main contribution to a major discussion that I hope to make here and hopefully have already made in other papers, it is to demolish this illusion of normalcy. Philosophy of logic in any reasonable sense is in a crisis. For one thing, what is the logic that "philosophy of logic" is supposed to be philosophy of? With due allowance to the complications and controversies connected with the notion of logic, in practice the basic part of logic everyone relies on is the quantification theory that goes back to Frege and that is known as (the received or traditional) first-order (FO) logic. It is also thought generally that for the kinds of human reasoning that go beyond traditional FO logic we have to resort to set theory. But in the way set theory is usually developed axiomatically, the logic used there is the same traditional FO logic.

The central role of traditional FO logic today is that various results that can be seen to establish its limitations are hailed as the greatest achievements of logic in the 20th century. These results include the incompleteness, undefinability and unprovablity results by Gödel, Tarski, Paul Cohen and others.

Yet this traditional FO logic originated by Frege was seriously inadequate right from the beginning. Mathematicians such as Weierstrass were already at Frege's time using an explicit although unformalized logic of quantifiers that is significantly richer than Frege's *Begriffsschrift* or his followers' traditional FO logic. Ever since Frege, philosophy of logic has in effect been philosophy of a wrong or at least flawed (too weak) logic.

This problematic situation is diagnosed in my paper, "Which mathematical logic is the logic of mathematics?" It turns out that Frege's mistake was not a casual one but due to an inadequate conception of the semantics of quantifiers. He overlooked the role of the formal dependence relations between quantifiers as indicating the material dependence relations between their variables.

The richer logic of mathematicians was rediscovered only in the nineties by Hintikka and Sandu and systematized under the misleading title "independence-friendly (IF) first-order logic." Meanwhile, no one had straightened out Frege's way of thinking. This led to the paradoxes of set theory and the entire "crisis of foundations" of mathematics (see my paper "IF logic, definitions and the Vicious Circle Principle".) Unfortunately, no one could give a correct diagnosis of the underlying mistake. The so-called Vicious Circle Principle was an attempt to do so, but its true meaning was misunderstood by virtually everybody.

3. What is the proper role of philosophy of logic in relation to other disciplines, and to other branches of philosophy?

Thus problems in the philosophy of logic lead directly to the development of analytic philosophy and indirectly to questions of its nature and prospects.

To give just one example, Frege's status as an expert in the philosophy of logic and in the philosophy of mathematics has to be re-evaluated. It has become embarrassingly clear that Frege not only did not have anything to do with cutting-edge work in mathematics (apart from logic); he had no idea of the conceptual problems and other issues arising out of that work. And asking for such awareness is not an unfair or anachronistic requirement. C.S. Peirce, unlike Frege, was not a professional mathematician and yet it turns out that he not only avoided Frege's mistake about the semantics of quantifiers but also understood fully the principles of reasoning that mathematicians like Weierstrass were using. Unfortunately, Peirce did not try to formalize them, as he could easily have done.

For another example, the significance of the incompleteness and undefinablility results mentioned above must be reassessed. They are of course technically correct, but limited in applicability. In the same sense in which Tarski showed that truth is not definable for a traditional FO language in the same language, truth can be so defined if we use FO IF logic instead of the received FO logic, as Tarski did. At this moment straightening out these problems is the most urgent aspect of the proper role of the philosophy of logic in philosophy at large.

Mathematicians tried to escape the paradoxes of set theory by developing set theory in the form of FO axiom system. Unfortunately, this was another catastrophic mistake. The most common such axiomatization, the Zermelo-Fraenkel (ZF) system, still violates the rightly understood Vicious Circle Principle. No wonder one can prove in it interpretationally false theorems, among them the existence of true propositions without a full set of Skolem functions.

More about the woes of FO set theories is found in my forthcoming paper, "Axiomatic set theory *in memoriam*". The crucial mistake was the use of FO axiomatization in set theory. The models of a FO axiom system are structures of particular objects. Yet what is studied in set theory are, of course, structures of sets. How a study of the former is supposed to inform us about the latter has never been explained. In the interest of historical accuracy, perhaps it should be remembered here that Zermelo never intended his axiom system to be first-order.

4. What have been the most significant advances in philosophy of logic?

Only time will tell what the most significant advances in the philosophy of logic have been. It is not difficult, however, to find a candidate for the role of the most significant regress in the philosophy of logic. It is the central role of the notion of logical system in the theory and the application of logic. This notion goes back to the very idea of symbolic logic. The idea is that logical deductions can be carried out in the same formal way as calculations. The notion of logical system carries out this idea by requiring that such a system consist of a finite number of formal axioms and a finite number of formal rules of inference. (The finiteness requirement can be weakened to recursive enumerability.)

This seems to be fine. Unfortunately, it turns out that in the light of results like Gödel's, restriction to a finite number of rules of inference is unrealistic. In mathematical or scientific practice, only trivial theories can be forced into the form of a "logical system".

This is not a significant limitation on the theory or the use of logic. In hindsight, it might be seen as an indication of the richness and power of nontrivial logical reasoning. To try to limit logical inference to a finite number of formal rules might seem to be due to a far too narrow a view of what logic is and what it can do.

Unfortunately, the damage has been done. A widespread view more or less identifies logical studies with studies of different possible "logical systems". Such a study is very important but rightly understood it is not about what logic is or can do, but about what can be done in logic purely mechanically. What results like Gödel's first incompleteness theorem deal with are not limitations on what logic, mathematics, axiomatic method or human reason can do. They are limitations on what computers can do in logic, not on what humans can do.

More generally, the alleged central role of "logical systems" has led to a misleading perspective on what logic is all about. It is not a study of systems of inferences, but in the last analysis a study of their semantical (model-theoretical) justification. It would be healthier to think of logic as the science of thought-experimentation with different models and model-constructions than as a study of systems of rules of inference. Overemphasis on inference has led logicians to neglect what may very well be the most common use of logic in our actual reasoning: not to show that something is necessary or follows necessarily, but to show how something is possible. The rules of logic can be thought of as accomplishing the one or the other in the sense that can be seen in operation for instance in the tableau method rules.

The "how possible" reasoning would be especially useful to discuss in the teaching of logic. For instance, most of the feats of "deduction" and "logic" in proverbially good reasoners like Sherlock Holmes are not arguments to show that something must have happened, but to show how it could have happened. They provide no end of non-trivial examples of "how possible" uses of logic.

But the "logical system" idea has corrupted logicians' thinking in a yet subtler way. If what happens in logic, including its applications, is in principle application of a given finite number of formal rules of inference, then all logical and mathematical problems become in principle computational problems. For instance, finding a proof for a complicated mathematical theorem will on such a view be like finding an algorithm that produces the different steps of argument. As most people's experience with computers and computation has shown them, such computational problems quickly become extremely complex, requiring in effect an industrial-size and style cooperative project for their solution. The "logical system" thinking has in this way led a surprisingly large number of logicians and mathematicians to a surprisingly deep-seated belief in the hopelessness of individual researchers' attempts to solve major problems in mathematical sciences.

This pessimism may be realistic as far as old, long-studied mathematical problems are concerned. But there is little reason to think that it applies in the same way to problems with a complex conceptual nature. I may be overly optimistic, but I do believe that correcting the mistakes logicians have been working under will open possibilities for major new results.

I have elaborated these points in my paper, "A scientific revolution in real time."

5. What are the most important open problems in philosophy of logic, and what are the prospects for progress?

The most important open task in the philosophy of logic is to get rid of the mistakes and misunderstandings of the past.

Bibliography

Hintikka, Jaakko, 1996, *The Principles of Mathematics Revisited*, Cambridge U.P., Cambridge.

Hintikka, Jaakko, 2012,"Which mathematical logic is the logic of mathematics?", *Logica Universalis* vol. 6, pp. 459-475

Hintikka, Jaakko, 2012, "IF logic, definitions and the Vicious Circle Principle", *Journal of Philosophical Logic* vol. 41, pp. 505-517

Hintikka, Jaakko, 2014, "A scientific revolution in real time", Teorema vol. 33, pp. 13-27

Hintikka, Jaakko, forthcoming, "Axiomtic set theory *in memoriam*"

12

Dale Jacquette

Senior Professorial Chair in Philosophy
Universität Bern, Abteilung Logik und theoretische Philosophie

---

1. Why were you initially drawn to the philosophy of logic?

I assume that my experience is not very different from that of many other researchers in the field.

I was directed toward standard elementary symbolic logic my first year pursuing an undergraduate philosophy major. Previously, I had read around a bit in several areas of philosophy, but I had not encountered formal logic as a discipline in its own right or as a rigorous instrument of conceptual and inferential analysis. I did not yet know that such a thing existed.

I was interested in arguments, criticism, and justification for philosophical positions. I knew that reasoning was supposed to have logical structural properties and to satisfy basic inferential requirements. What I did not know was that there is an intrinsically interesting game-like formal theory of logical relations. Such appreciation came only later. The first challenge was to master basic techniques, translation skills, decision algorithms, natural deduction rules, and proof strategies, including structural dissection of selected philosophical arguments.

That was the first semester course, which I suppose was the usual first course in symbolic logic at the time. A follow-up course the second semester took me through the fundamentals of metatheory, semantic valuations, syntactical and semantic Henkin-consistency, completeness and compactness proofs, and deductive incompleteness in the classical limiting metatheorems. Juicy domain comprehension and cardinality results were still in the future.

I did well enough in this first exposure to symbolic logic to be invited by my professor to assist and tutor fellow undergraduates my second year of college. By that time I had identified analytic philosophy as my main interest and focus of my study, which is not necessarily to say *mainstream* analytic philosophy. If such a thing even exists, I should probably want to resist it on general contrarian principles to learn what happens when its principles are challenged.

The historical connection between mathematical logic and the philosophical foundations of mathematics was inescapable. The more I learned about the one, the more I understood the importance of learning about the other. These topics, as any informed reader will already know, were originally motivated by philosophical problems and a desire to arrive at a philosophically circumspect groundwork for all of mathematics, a correct explanation of the nature of mathematical objects, logical inference, the modality of logical and mathematical truth and proof, and the relation between logic and mathematics, among many others. This is only to say that it was my increasing interest in logic that fueled my growing preoccupation with the philosophy of logic, rather than the other way around.

Exploring logic as its own subject, and my interests in analytic philosophy, led me to independent inquiry in the philosophy of logic. I was not guided toward logic as an instrumental means to another end, or as an incidental sideline, but as an end in itself that turned out to raise fascinating philosophical questions while providing the formal language for their exact expression. That is the point at which I became interested specifically in philosophy of logic.

Eventually my work in logic also became a useful tool, one among complementary interests in philosophy, and something more. The trajectory, accordingly, in my particular case, as my interests and skill sets developed, was not from philosophy to philosophy of logic, despite the fact that I was also working in parallel on a variety of questions in several areas of philosophy at the time, but rather from logic to philosophy of logic within a broader network of reading, thinking, and coursework in analytic philosophy. Logic opened itself up as a self-contained world as I came to learn more about it, and that increased my mounting curiosity about an array of systematically related issues in the philosophy of logic.

I became involved in philosophy of logic, consequently, because of the importance of logic in my philosophical education. As I gained confidence, during my graduate program in philosophy, logic was always at the center. An engagement with many issues in the philosophy of logic kept pace in a symbiotic synergistic interaction with my interests in formalisms. In retrospect, these studies have progressed in an entirely natural, almost inevitable way, with a sense of fit and rightness about it that takes its own direction in my orientation, energy and passion for logic and philosophy.

2. What are your main contributions to the philosophy of logic?

The subjects in philosophy of logic to which I was drawn in my dissertation research and continuing throughout my later work in logic and philosophy included most prominently the relation of reference, predication and quantifier interpretation in the functional or predicate calculus. I was dissatisfied with extensional interpretations of naming, and of the true predication of properties and quantification generally in scientific discourse and the history of science, which I found strewn with at least as many meaningful philosophically interesting falsehoods as truths.

What are we to say of distinct false scientific theories or historical explanations that, like works of fiction, appear to posit distinct nonexistent objects whose corresponding predicates have identically null extension? If such statements have meaning, then they must minimally refer to distinct nonexistent objects, which no extensional semantics allows. A theory of phlogiston is not about the same objects as a theory of vortices, even though as nonexistents they are logically indistinguishable in what was then and to a large extent is still today a conventional extensionalist *existence-presuppositional* reference domain and correlative interpretation of truth conditions for universal and existential quantifications.

I sought a fully general philosophical semantics in the philosophy of logic when I first entered the field that would apply indiscriminately to true and false theories. I continue to think that the purity of logic demands an ontically neutral language for reasoning about true and false propositions alike. I wanted a semantics for existent concrete physical objects as well as nonexistent idealizations like the ideal gas, perfect fulcrum, frictionless surface, false hypothetical intended objects as phlogiston and vortices, the planet Vulcan again, and other *irrealia*. These abound in unpredetermined numbers in all areas of thought – including perhaps the most cherished and respected of contemporary scientific theories, which, despite wide acceptance and positive justification, could ultimately turn out to be false.

As we know from the history of the sciences, educated opinion about what exists can change despite being sociologically well-entrenched, in the course of a sufficiently powerful scientific or philosophical revolution. Among other areas of propositional expression, like myth, fiction, and religion (most or possibly all of it), science is subject to inadvertent reference to nonexistent objects. It has been my argument that an adequate universally applicable logic needs to accommodate reference and true predication to both existent and nonexistent objects. Logic itself knows nothing of what happens to exist or not exist, as I sometimes emphasize, so that the meaningfulness of random propositions cannot depend on the contingencies of actual existence and nonexistence. I also understood that reasoning about nonexistent objects is the key to understanding imagination, invention, and creativity in all areas of human activity.

Working out a logic for nonexistent objects was the first substantive problem in the philosophy of logic to which I devoted attention. Nor was I alone in the effort. A number of logicians had independently come to the same conclusion at roughly the same time, and, while still in a minority, we were all trying to develop logics and formal semantics to support reference and true predication of constitutive properties to nonexistent objects generally, in any field of discourse.

Whether what I have attempted to accomplish in philosophy of logic represents a *contribution* is probably best left for others to judge. From my perspective, I would say that what I have proposed in the philosophy of logic falls into two main categories: (1) critiques of received logical positions, assumptions, presuppositions and conclusions, involving the presentation of new paradoxes; (2) development of positive alternative logical structures, semantic principles, predication and inference rules, reference domain comprehension and cardinality results, along with philosophical applications in understanding dialectic at a variety of levels in the general field of argumentation studies.

Specifically, in category (1) I would cite the following:

- Formal criticisms of extensionalism, and of Quinean criteria of ontic commitment.

- Formalization of a nonidentity paradox and generalization puzzle to expose previously underappreciated counterintuitive expressive limitations of standard extensional predicate-quantificational logic.

- Argument for rejecting the liar 'paradox' on three distinct but converging grounds as not genuinely paradoxical.

- Rejection of Russell's 'paradox' as either inadequate to support the inference that if the Russell set as the set of *all* sets that are not members of themselves is a member of itself, then it is not a member of itself; or, like the barber 'paradox', implies only the nonexistence of the Russell set of *all and only* those sets that are not members of themselves.

- Discovery of the soundness paradox, that for argument (S): S is unsound; therefore, S is unsound.

- In category (2), with more positive results, I would emphasize the following.

- Development of an intensional logic of nonexistent objects.

- Analysis of formal diagonalization techniques in symbolic logic and the foundations of mathematics.

- Proof that truth-breakers are needed alongside truth-makers in an adequate truth conditions theory.
- Proof that Grelling's heterological paradox is recoverable despite simple type theory restrictions.
- Proof of the intensionality of identity, and of truth functionality.
- Arguments in support of the finite but inexhaustible cardinality of reference domains.
- Demonstration that, contrary to recent objections, Wittgenstein's joint negation operator in the general form of proposition in *Tractatus Logico-Philosophicus* is sufficient within picture theory constraints to comprehend all quantified logical expressions, existential and universal.
- Argument in deontic logic to establish formal limitations on the generalizability of Kant's principle that 'ought' implies 'can' in theoretical ethics.
- Argument to establish formal limits of Gödel incompleteness results in classical metatheory.
- Argument to limit applicability of Cantor diagonalization in arguments to establish the cardinality of irrational real numbers.
- Argument to show by computable Cantor-style diagonalizations that Turing machine computable real number sequences are non-denumerably infinite in cardinality.
- Argument for deductivism, reducing the so-called informal or rhetorical fallacies to forms of distinct deductive invalid inferences.
- Logical foundations for collective intentionality in intensional semantics.

I have also written on selected historical topics in philosophy of logic, including commentary on, among others, Aristotle, William of Sherwood, John Burleigh, John Buridan, Franz Brentano, George Boole, Charles Sanders Peirce, Gottlob Frege, Bertrand Russell, G.E. Moore, Ludwig Wittgenstein, Edmund Husserl, Rudolf Carnap, W.V.O. Quine, Josef Bochenski, and Saul A. Kripke.

My writings, monographs, edited works, essays in journals, and book chapters, in philosophy of logic as in other research areas, are catalogued, often within a year of publication, but almost never fully up-to-date, online at: http://www.philosophie.unibe.ch/content/ueber_uns/team/mitarbeitende/jacquette/publications/index_ger.html.

I recommend for general topics in philosophy of logic, Jacquette, *Meinongian Logic: The Semantics of Existence and Nonexistence* (Walter de Gruyter 1996), *Ontology* (McGill-Queen's University Press 2002), and *Logic and How it Gets That Way* (Acumen 2010), as well as my editorial comments in my textbook, *Symbolic Logic* (Wadsworth 2001), and in collected works, *A Companion to Philosophical Logic* (Blackwell 2002), *Philosophy of Logic: An Anthology* (Blackwell 2002), and *Philosophy of Logic* (Elsevier 2007). A collection of my essays edited as chapters on Meinong's philosophy since the time of publishing *Meinongian Logic*, is now contracted with Springer Verlag for (2015), with respect but no apologies to Martin Heidegger, under the title, *Alexius Meinong: The Shepherd of Non-Being*.

3. What is the proper role of philosophy of logic in relation to the other disciplines, and to other branches of philosophy?

The first issue that needs to be settled is whether logic itself is a branch of philosophy. I think that while philosophy of logic is manifestly a branch of philosophy, logic plays a complex role both in and out of the general study of philosophy, and more especially in the philosophy of logic.

It is standard to treat logic either as ancillary or foundational to philosophy. After all, philosophy involves reasoning and argument at every level, and logic is in some sense the theory of good reasoning and especially of logical inference in argument. We need logical principles for proper inference in argument contexts in philosophy. True enough, but we equally need logic in every other discipline, all of which in one way or another also depend on correct reasoning in order to advance their conclusions.

This fact suggests that logic stands outside of all other fields, including philosophy and the philosophy of logic. To a certain extent this seems obviously right, with implications also for the question of how philosophy of logic relates more generally to philosophical inquiry. Where history, law, the special sciences and everyday reasoning are concerned, wherever judgment is reached by argument and justification rather than unsupported intuition, revelation or authority, logic is needed for the acceptance of propositions. Wherever, one might say, propositions and the relations of their truth values provide the basis for thinking, logic is present and correct logical principles of reasoning are at work. The situation in philosophy of logic is nevertheless intrinsically more complex, because in that case we are using logic to reason about logic. This can easily be made to appear as bootstrapping (paradoxical in a self-referential or self-justifying way), or as an instance of hermeneutic circularity (of an innocuous or vicious type, depending on one's point of view).

The dilemmas that arise are typified by the following. If we are dissatisfied with classical logic CL, then we may want to replace it with a nonclassical logic NCL1, NCL2, etc. If we rely intuitively on CL to support the development of any of these nonclassical logics, then it must seem that of the two, CL is the real, fundamental, or more deeply underlying logic. The undergirding of NCL1 by CL or in support of NCL2 can be seen as showing that nonclassical logics are not indispensable to all reasoning, but are derivative from NCL1, NCL2, etc. are not as basic to reasoning generally as CL.

There is a great difference between supplementing CL with notations for modalities, times, places, persons, actions, or for imperatives and interrogatives, and many other things whose logical structures are interesting or philosophically essential to consider, and replacing CL with an NCL that is logically inconsistent with CL (as when NCL logically implies for all propositions p, that p and not-p is true). If we try to use a nonclassical logic NCL2 in support of the original nonclassical logic NCL1, then there is no independent logical ground for the nonclassical formalism. In effect, there is no dialectical motivation for turning away from rather than supplementing CL. It is simply presupposed from the outset, rather than established by reasoning in response to problems, paradoxes, and the other sorts of cognitive challenges for which a logical language is usually invented or devised, that CL is too limited for expressive purposes, and needs to be replaced. The possibilities of building on CL are often overlooked nowadays, when the trend is more toward replacing CL at the drop of a hat with something interestingly extravagant.

To argue in favor of CL by means of CL is viciously circular, and to argue in favor of CL by means of an NCL is unsound. To develop an NCL1 from CL is not only unsound but deductively invalid, whereas to develop NCL1 from NCL1 itself is as viciously circular as to argue in favor of CL by means of CL. To develop NCL1 from another NCL2 only pushes the problem back one further step, as we ask how NCL2 is uniquely described and its expressive and inferential advantages justified. How does NCL2 relate to CL in supporting NCL1?

We cannot proceed in philosophy of logic without good answers to these fundamental theory-building metalogical metaphilosophical questions necessarily involving or relating a new NCL formalism to CL. CL is always available for building rather than razing and replacing. The choice is that of theorizers, theory-builders, in the philosophy of logic. If we supplant CL out of more than idle curiosity, with some NCL, then we must try to justify the decision in terms of what we believe we need our best logics to do. The formalization of propositional expression and the licensing of inferential relations primarily involves indefinitely large sets of semantically pre-evaluated propositions, sentential or abstract, under any combination of univocally assigned truth values. Are we to preserve CL and set it alongside all the NCLs? Or do we demolish CL in order to make room for preferred NCLs? The philosophy of logic is called upon precisely to answer these kinds of fundamental questions that may be vital to symbolic logic's development and its philosophical approval rating.

What, then, to do? How do we surmount this two-edged objection? Here we encounter a difficulty in the philosophy of logic that has seldom been remarked or investigated, let alone resolved. The existence of the problem already marks a significant difference in the relation between logic and other extra-logical and even extra-philosophical disciplines, as compared with the relation between one logic and another logic, or between logic and philosophy, with logic considered as a part, or even the most fundamental underlying foundation, of philosophy. It is not good enough in trying to avoid the dilemma merely to say that logic is discovered rather than invented. That distinction also touches deeply on one of the most important and troublesome choices in the philosophy of logic. It encourages logicians to conceptualize the purpose of logical formalization in abstract terms either as altogether independent of natural reasoning, or as regimenting the reasoning patterns that actually prevail at the root of all inference at its best. Whatever its ultimate purpose, one can say that if a formal language does not map satisfactorily onto a large body of meaningful discourse, then we should not first think that there is anything wrong with the meanings that language and other made objects are used to express. We should look first to the logic and inquire into its adequacy. It is by such pressures laid upon a classical logic that variant formalisms have developed in response to particular analytical challenges.

Complete with syntax, formal semantics, axioms and inference rules, or their natural deduction or graphic equivalents, CL and any NCL must earn their keep by providing satisfactory logical structures for the symbolization of individual logically complex expressions, and for what are considered to be correct inference structures. The intention is in this way to include CL and NCLs alike. I take the meaning of any expression to be the propositional content intended by a subject in predicating a property of an object. Meaning in this sense is reflected in a clear understanding of the propositional content of any predication of any property to any object. Regardless of how thought and a natural language or logical notation expresses predications, we speak in this connection specifically of propositional meaning, where the meaning of an expression can in principle be correctly cashed out as a proposition. The same is true without prejudice, then, when logical reasoning is turned inwardly upon logic itself, just as when logic is applied in the sense of being satisfactorily mapped upon any other subject. The logic of logic, the logic that underwrites a given exposition of logic or theory or philosophy of logic, is itself not yet philosophy of logic. A philosophy of logic must say something contentful concerning the nature of logic and logical relations, and there are many things to say. Logic, even the logic of a philosophy of logic, does not recommend reference domain comprehension principles for a first-order CL (or NCL). Satisfactory semantic connections beyond possible one-one correspondences between logic and meaningful expressions can obviously only be arrived at through a course of philosophical inquiry, and arrived at only as a specifically philosophical, rather than purely logical, conclusion. The same is true more generally of philosophy, assuming logic or philosophy of logic to be some kind of part or component of a complete broadly scientific philosophy. We test logic in applications other than logic first. When a formalism comes up shining, we may come to trust it also in thinking about logic itself as just another application in its broad range to draw upon in developing, among other things, a philosophy of logic.

Such a proposal, promising as it may at first seem, even if construed as a last straw for philosophy of logic, cannot imaginably suffice. Systematized as a formal language or not, the conflict posed by the dilemma at bottom concerns any logic, regardless of how it is instantiated or expressed, required for whatever reasoning is supposed to govern thinking about logic. This includes the logic in question itself, in developing a philosophy of logic. The problem, so fundamental to logic, philosophy, and the philosophy of logic, is stubbornly persistent and not easily or obviously resolved – if in the end it is solvable satisfactorily at all. One possible answer is to suppose that logic is not after all a part of philosophy, but in some sense external to or excluded from it. If so, then we can speak without circularity of logic as reflected theoretically in all disciplines, even in thinking about logic. We can choose in a principled way from among alternative preferred formalizations of logic, on the one hand, and, on the other, the applications of a preferred logic in other disciplines, including philosophy more generally, and the philosophy of logic in particular. Logic is then its own logically independent thing. It is foundational to every discipline including philosophy, mathematics, and hence, *a fortiori,* to philosophy of logic and philosophy of mathematics, not to mention philosophy of language.

For this reason alone, the logicist program of trying rigorously to reduce mathematics to principles of pure logic should have been understood from the outset as hopeless. The task was misconceived, and logicism developed as a reduction of elementary mathematics to the principles of specific applied logics. Mathematical concepts, unfortunately, are there to be found among the arid logics available in formalizations, only as applied logical terminologies that are as close to pure logic as F = ma in applied mathematics of kinematics is to 1+1 = 2 in pure elementary arithmetic. Later knockdown objections to logicism, such as Gödel's incompleteness proof, are best interpreted as corroboration for an effort that was logically, conceptually, and cognitively doomed to failure anyway, and whose undertaking would have been most wisely avoided in the first place, except perhaps as a philosophical experiment confirming exactly the predictable calamity.

Logic, then, is not systematically part of philosophy, even though the study of logic and its applications may be of special philosophical interest and importance. The study of logic contributes indirectly, in somewhat the way that empirical discoveries in the natural sciences often nourish and positively influence philosophical reflection in all fields. The fact that we speak intelligibly of philosophy of logic, as we do of philosophy of mathematics and philosophy of science, where mathematics and science are not generally considered part of philosophy, confirms the distinction. Whereas, in contrast, we do not typically recognize as legitimate accepted categories – at least I have never seen a book or professional essay titled – "The Philosophy of Ethics", "The Philosophy of Metaphysics", or "The Philosophy of Epistemology", as opposed to simply, ethics, metaphysics, or epistemology, as subdisciplines of philosophy. We should accordingly already be alerted to the fact that logic is not part of philosophy by the fact, in contrast, that we *do* speak of philosophy of logic, just as we speak of philosophy of language, philosophy of mathematics, philosophy of action, philosophy of literature, and the like.

The first principle in a philosophy of logic on which to agree might therefore be that logic is vitally important to, but not strictly a part of, philosophy. Logic must be independent of philosophy in order for there to be a noncircular philosophy of logic. An adequate philosophy of logic investigates all philosophically interesting aspects of logic, and in the process presupposes and puts to use specific syntactically formalizable expressive and inferential logical principles. With anything less substantial on board there is in the first place no logic in the proper sense of the word about which to philosophize. How to pull off this balancing act of using logic to develop a philosophy of logic without slipping into vicious circularity or deductive fallacy is one of the main challenges for a methodologically circumspect philosophy of logic.

The best solution is (perhaps) first to develop as ontically neutral a base logic as possible, as an extension of CL, which might be considered an NCL. Second, to minimize the use and hence the ontic impact of CL or its specific NCL extension in developing the arguments necessary to support all the usual commitments, or their opposites, of an independently appealing philosophy of logic. The less we rely on a specific logic, CL or NCL, to make progress in the philosophy of logic, the less likely we are to smuggle specific logical preconceptions surreptitiously into our most general thinking about the nature of logic, its scope and limits.

4. What have been the most significant advances in the philosophy of logic?

Assuming I am not asked to explain why I find any of these results significant, I would include in the list some of the most noteworthy pioneering philosophical efforts to determine the correct referential domain contents and cardinality of a semantically complete predicate-quantificational logic, philosophical justification for an adequate comprehension principle in formal and intuitive semantics, the scope and limits of extensional versus intensional interpretations of basic logical relations, type-theory, Gödel-Church-Rosser and Löwenheim-Skolem proofs, formal logical and semantic paradoxes, the wide variety of discoveries made in the course of trying to solve, forestall, or avoid antinomies, nonclassical logics of nonexistent objects, modal formalisms and explorations of the meaning of modal propositions in terms of logically possible worlds and related (especially set theoretical) structures, relevance logics of entailment, nonmonotonic, dialethic and paraconsistent formalisms, formal results concerning the relation between deductive and inductive reasoning, theories of universals, formally interesting reduction strategies (however ultimately futile, as previously mentioned), of mathematics to logic, automated proof discovery, and instructive applications of logic in a wide variety of conceptual analyses.

Philosophical responses to the most significant advances in logic are generally among the most significant advances in philosophy of logic. The most important questions for a philosophy of logic are therefore how to make sense of these discoveries, and what philosophical conclusions we can responsibly draw from new work in formal symbolic logic.

For example, I find many responses to the liar paradox logically very ingenious, beginning with Alfred Tarski's hierarchies of formal metalanguage truth operators among the formal conditions for true sentences in an object- or lower-level meta-language, and including dialethic or paraconsistent logics, along with many other strategies. Aside from its often independently interesting formal structures, I nevertheless question the value of much of this work when there are arguments to show that the liar itself is not a genuine paradox in the first place. It appears that only one of the dilemma horns, from true to false, goes through, and not the partner horn, from false to true. We are generally assured when the liar paradox dilemma starts to get off the ground, that the liar sentence is either true or false. If the liar sentence is true, then the inference from the liar sentence to a necessarily false paradox must be deductively invalid. If the liar sentence is false, on the other hand, then it must be logically false, from which any conclusions whatsoever follow trivially, under the so-called paradoxes of strict implication. The liar paradox is supposed to embarrass CL, but the paradox inference cannot be logically supported except as deductively trivial by CL. If we adhere to the limits of CL, then the liar paradox cannot go through as a deductively valid or deductively trivial inference from the liar sentence. There is, to my knowledge, no logically interesting derivation of the liar paradox from any true or false liar sentence.

This is similar to many other areas of logic where starting places are repeatedly taken for granted as something to build upon, rather than subjected to more searching philosophical scrutiny.

5. What are the most important open problems in philosophy of logic, and what are the prospects for progress?

The glib, but in some sense irrefutable answer to this question, is that all problems in the philosophy of logic are open. Else, they would not be problems. The most important of these problems in philosophy of logic, from my perspective, are the perennial challenges of understanding the nature, scope and limits of logic. How is it that there are repeatable patterns of reasoning, some deductively valid and others invalid? What is validity and what is invalidity? Are these relations objective and mind-independent, or do they supervene on even more fundamental cognitive structures at the basis of thought? Do they merely reflect the contingencies of how our brains are wired, such that we cannot imagine any other possibilities? Can we make sense of the idea that there is or could be a universal logic, as some theorists have suggested? Or is logic more correctly considered to be a family resemblance concept, with many different kinds of logics like the families of language games that the later Wittgenstein describes? What are propositions? Are they abstract and universal, or merely sentence types of a certain kind that propose an object or objects having a certain quality, or standing to one another in a given relation? What is the exact modality by which the conclusion of a deductively valid inference is supposed to follow necessarily from its assumptions? How does logic relate to psychology? Are they independent of one another, or is logic, as the theory of good and bad reasoning, inextricably interconnected with and ultimately an expression of certain anthropological psychological facts about the biological and cultural evolution of human intelligence?

As to the prospects of progress in these areas, I think that, in view of the perennial nature of the problems, they are in one sense probably about the same as they have ever been. The fact that there are more laborers toiling in the vineyards of logical studies today than ever before, standing on the shoulders of impressive achievements in the last two centuries, and with powerful digital processing tools on desktops worldwide, cannot be ignored. There is cause for optimism about the likelihood of achieving greater and more lasting profound discoveries, using computer modeling technology and automated proof algorithms as improved devices for both investigating and rapid global communication of new results, encouraging more lively productive critical interactions. There are more researchers. There are more formalisms with different solutions to similar problems from which to choose and about which to argue philosophically. If we are to judge by the recent history of the subject, then we should also expect an interesting equally accelerated floruit of new problems in and for logic and the philosophy of logic.

The sheer number of researchers and their increasing technical competences and sophistication, opportunities for cross-fertilization of ideas and valuable feedback, insofar as these are specifically philosophically motivated, increase the opportunities for conflict, disagreement, and dialectical opposition as to what is or is not a problem of logic, and what is or is not a solution. That diversity of a greater labor force might also be seen as an obstacle or impediment to progress, in the sense of attaining a convergence of informed opinion laying the groundwork for further pursuit of the most persistent urgent questions in the philosophy of logic. The state of the art at any chosen historical point will no doubt be an amalgam of these and related elements for evaluating progress in the enterprise, of splinterings within the general field of philosophy of logic into ever more finely subdivided ideological and methodological camps. The health of the subject depends on just such continually renewed dialectical oppositions and refinements on multiple sides of every dispute. In light of increasingly sophisticated formal logics, there must be progressively sophisticated philosophical questions, more adaptive styles of argument, formal distinctions, definitions, principles, axioms, inference rules, and philosophical, ontic, metaphysical and metalogical responses to each innovation and considered application.

13

Penelope Maddy

Distinguished Professor of Logic and Philosophy of Science

Department of Logic and Philosophy of Science, University of California, Irvine

---

1. Why were you initially drawn to the philosophy of logic?

I got into philosophy, coming from mathematics, because I hoped to find a viable philosophical setting for the view that the Continuum Hypothesis (CH) has a determinate truth value. In time, some of the assumptions (then widely shared) that underpinned this 'set-theoretic realism' (like the indispensability arguments) began to trouble me and eventually I lost my confidence that there's a fact of the matter about CH waiting to be discovered. Still, it didn't seem right to say that anything goes, that one consistent set theory is as good as any other, so my attention turned to other ways, less straightforwardly metaphysical ways, to give substance to these impressions.

Now any philosopher who suggests that perhaps mathematics isn't a descriptive practice, that its claims aren't properly true but instead enjoy other virtues -- any such philosopher is almost certain to hear the incredulous rejoinder: so you mean to say that 2+2 isn't really 4?! It isn't 'so you mean to say that there aren't really more real numbers than natural numbers?!' or '... that there aren't really non- separable metric spaces?!'; it's always about '2+2=4' (or '7+5=12' for Kantians). What would I say to that?!

Truth be told, this always felt to me like sleight of hand: where I wanted to be talking about higher set theory, the objector shifted ground to arithmetic, and not to Fermat's Last Theorem or anything like that, but to elementary arithmetic, to simple identities between small numbers. And this was undeniably an effective rhetorical move: who would be so silly as to think 2+2 isn't 4?!

My suspicion was that these elementary arithmetical facts were really more like ordinary logic than they were like higher set theory, that there was no reason to think an account of the latter should also hold for the former, but to spell this out would require an account of logical truth. By this route, I landed in the philosophy of logic.

2. What are your main contributions to the philosophy of logic?

My 'contributions', such as they are, take place within a setting so resolutely naturalistic that many would say they come to nothing, they don't truly engage with the philosophy of logic at all. Even from this 'second-philosophical' point of view, I don't address what many regard as the central question: what separates logic from non-logic? What I do instead is consider some examples that everyone would take to be logical -- e.g., 'if the book is either red or green, and it's not red, then it must be green' -- and ask what grounds them, what makes the corresponding inferences reliable? My concern isn't what makes them logical, but what makes them true (or valid).

The Second Philosopher's answer is that a rudimentary part of logic, including these straightforward examples, rests on very general features of the worldly situations they describe (roughly their consisting of individual objects with properties, standing in relations, with dependencies holding between some such situations and others). These features are contingent, and for that matter, aren't present in all aspects of our actual world (e.g., the micro-world), but where they are present, rudimentary logic is reliable. On this account, logic isn't necessary or a priori (except perhaps in a weak externalist sense), and even if it's analytic in one way or another, this isn't enough to guarantee its correctness. The result is a robustly realistic account of rudimentary logic, entirely in line with Russell's remark (in *Introduction to Mathematical Philosophy*) that 'logic is concerned with the real world just as truly as zoology, though with its more abstract and general features' (Russell may well have meant this less straightforwardly than I do).

But rudimentary logic falls short of full classical logic. The move from one to the other involves a series of idealizations, introduced, as idealizations are, to generate a simpler, more powerful theory, and justified, as idealizations are, to the extent that they're beneficial and benign. For comparison, I agree with Timothy Williamson that the various proposals for a 'logic of vagueness' are unwieldy and that classical logic is much to be preferred, but he achieves this happy outcome at the price of epistemicism (there's a fact of the matter about whether or not Joe is bald, it's just that we can't, in principle, know what it is), an unattractive move that strikes me as both unacceptable and unnecessary. The Second Philosopher takes an alternative route to the same goal, acknowledging the obvious fact of indeterminacy, but retaining classical logic with the law of excluded middle as one of its key idealizations. It's clearly beneficial, as Williamson emphasizes, and it's also benign -- as long as the resulting instrument is used with care, for example, as long as we don't insist on long chains of sorites reasoning. After all, any idealization will run you into trouble if you abuse it!

3. What is the proper role of philosophy of logic in relation to other disciplines, and to other branches of philosophy?

As indicated above, it seems to me difficult to face the metaphysical and epistemological challenges of the philosophy of mathematics without a clear view of the philosophy of logic.

4. What have been the most significant advances in the philosophy of logic?

If I had to pick one, I'd go with Church's theorem, the discovery that there's no mechanical procedure for determining whether or not a given first-order sentence is valid. The epistemic transparency of propositional logic doesn't extend to the language of predicates and quantifiers. It's stunning!

5. What are the most important open problems in philosophy of logic, and what are the prospects for progress?

The questions that spring to mind may not sound like philosophy of logic at all, and maybe they aren't, but I'd like to know more about the logical understanding of infants and young children. Here are a couple of examples ...

In famous and ground-breaking experiments, pre-linguistic infants apparently appreciate, for example, that 1+1 is 2. This seems to me to involve rudimentary logic, and I hope that at some point some of these gifted experimentalists will address related questions from this perspective. For example, what sorts of inferences (apart from the 1+1=2 inference) are pre-linguistic infants capable of drawing?; do they grasp something like a universal quantifier?; and so on.

Also, much fascinating and important work has been done teasing out exactly how young children learn to recite the sequence of number words, then slowly come to understand that counting reveals the number property of the group counted. Eventually, these children also come to believe that there is no end to the sequence of numerical expressions and no largest number. How does this second transition come about? I'm tempted to think it marks the move from 'logic' to 'mathematics', at least in one way these terms are naturally used, but however that may be, I'd like to know more about how it takes place.

14

Lawrence S. Moss

Director, Program in Pure and Applied Logic, Professor of Mathematics and Adjunct Professor: Computer Science, Informatics, Linguistics, Philosophy at Indiana University, USA

I feel like a bit of a fake appearing in this volume: I'm not a philosopher of logic in any real sense. I'm a logician who has more than a little interest in philosophy of logic, and I'd like to think that some of what I have to say would be interesting to people in the field. When it comes to some of the core topics in philosophy of logic, topics like the analytic-synthetic distinction, the nature of truth, the relation of logic and metaphysics, and philosophical work on logic itself, I'm more of an "interested outsider" than an insider.

At the same time, I do feel more involved with philosophical logic, so I'll bend the topic of this volume. I know that it's not always so clear what the division of labor between philosophy of logic and philosophical logic is. When I started thinking about this interview, I went to the Routledge Encyclopedia of Philosophy. Graeme Forbes' entry there has two sections, one on modal logic, and the other on logic and language. These are topics that I'm rather interested in, and so I suspect that I belong here, or at least partially fit in.

My original training was in mathematics and logic. In fact, I was an undergraduate at UCLA, and in my first year I started in on abstract algebra, linguistics, computer science, and logic – all topics that I'm still involved with to this day. (So much for big changes!) The logic class was taught from Kalish and Montague's book *Logic: Techniques of Formal Reasoning*. (The third author of a later edition, Gary Mar, was a TA for one of my classes.) My major was mathematics, and indeed I stayed on at UCLA to get a PhD in Math, specializing in logic. My advisor was Yiannis Moschovakis, and surely he is a big influence on how I think about quite a number of things, including philosophical topics. But I frequented classes in several other areas, especially linguistics, and after finishing I spent a year as a post-doc at Stanford's Center for the Study of Language and Information.

It was first at CSLI that I was exposed to philosophy of logic as a living subject, and philosophical logic as well. I remember discussions of semantics, AI, situation theory, and much more. I sat in on Peter Aczel's course on non-well-founded sets; eventually Jon Barwise (then director of CSLI) and I would write a book on that topic. I think I had intellectually "graduated from" the world of mathematical logic that I had been raised in, and was more taken by areas of logic closer to computer science, philosophy, and linguistics. At some point, I noticed that I was receiving announcements of meetings on the many topics I was interested in, and occasionally invitations, but when it came to mathematical logic, I would read a report after it happened. After further post-docs at Michigan and at the IBM T.J. Watson Research Center, I found myself needing a job.

Jon Barwise was moving to Indiana University, I applied and got the job, and I've been here ever since. My position is in Mathematics, but I am even more of a fake for mathematics than for philosophy of logic. Jon was comfortable in many disciplines, and he was a ground-breaking contributor to several of them. He certainly was a model for me of how to be a multi-disciplinary academic. He himself believed (following advice of Abraham Robinson) that every five years, one should either change their job or their area of work. After I got tenure, he hinted to me that I should be more free to pursue more fundamental topics.

When Jon passed away in 2000, I found myself somewhat alone. I had been fortunate to work with a lot of people who thought about both their own subjects and also about more philosophical concerns, people like Ed Keenan, Joseph Goguen, Yuri Gurevich, Yiannis Moschovakis, and Rohit Parikh. I was so much younger than them; I still feel like their student in many ways. But by 2000, I was no longer the "youngest person in the room" as I once felt. I went through a time of thinking what would come next. I had to think more "politically" what it would mean to do multi-disciplinary work in logic in a large university. Coincidentally, I had to write a statement for the university about my work in a general way. I ended up writing "Applied Logic: a Manifesto" for this purpose, and at some later time I published it and also put it on my web site. Over the years, a number of people have commented on the Manifesto and felt influenced by it. Maybe one or two students a year write to me out of the blue saying that they appreciate it. One person told me that in the USA, what I have to say would be regarded as obviously false by mainstream logicians, but in Europe, it would be regarded as obviously true.

For this book, it might be useful to recall the themes of the Manifesto:

1. Applied logic is the application of logic and mathematics to human reasoning. At the same time, applied logic is connected with a host of topics in computer science and other areas. It should have connections to cognitive science and artificial intelligence.

2. Applied logic extends the traditional boundaries of logic to topics like *change*, *uncertainty*, and *communication*.

3. The division of logic into "mathematical logic" and "philosophical logic" is somewhat useful. But it is orthogonal to the division between "pure" and "applied" logic. When fitting into these categories, I see myself as mainly doing applied philosophical logic.

Another point worth quoting at length for this interview:

> Applied philosophical logic = theoretical artificial intelligence(AI). The slogan here is perhaps a bit of an overstatement, but the point is that work on the theoretical questions in artificial intelligence often looks back at earlier discussions in philosophical logic. One area where this happens is in the study of knowledge…Another is the study of *context*: how is it to be represented, and what role does it play in reasoning? If one wants to build a robot and make it rational, then the hard problem of deciding what *rationality* means will lead back to the parallel philosophical literature.

This pertains to two of the interview questions for this volume:

3. What is the proper role of philosophy of logic in relation to other disciplines, and to other branches of philosophy?

5. What are the most important open problems in philosophy of logic, and what are the prospects for progress?

Here is the point. Many of the hard questions in philosophical logic, and in philosophy of logic, happen to be related to questions in theoretical AI. Not all of them, but many of them. To mention a very contemporary example, in 2011, IBM's software system called Watson competed in the television game show *Jeopardy!* and won against two top-ranked human competitors. Certainly there is a lot of grist for the philosophical mill in this development: Whether or not Watson "understands" anything is up for debate. Watson's architecture is not at all what one would imagine at first.

I'm told by Wlodek Zadrozny, one of the people who worked on it, that Watson suggests a 'dialogue-like nature of language meaning', whereby meanings emerge via paraphrase and evaluation according to multiple criteria at the same time. In fact, the generation of possible answers is one of the things that enables Watson to succeed. As a logician, I'm curious that the amount of straight-out deduction involved in Watson is minimal. What happened to those thousands of years of work on reasoning? Returning to the questions above, my point is that work in contemporary AI (and related fields) really could inform philosophical discussion. At the very least, it could provide new viewpoints and problems. Conversely, the philosophical literature could well provide tools and viewpoints that could be useful in the "real world."

Another of the interview questions pertains to my "main contributions to the philosophy of logic." I would like to mention one area here, that of *natural logic*. Eventually, I would like to rework aspects of natural language based on computational linguistics. This parallels the points I made about AI above (but I am not working on those topics). At this time, I am trying to think about the notion of *entailment*; I suspect that some of the issues in this area will become interesting in philosophical logic.

Textbooks on model theoretic semantics often say that the goal of the semantics is to study entailment. (I happen to think this is not what semantics is really about, because aside from those first few pages of textbooks, there is very little about entailment in semantics.) I am trying to build *complete decidable* fragments of language. "Fragment" here means "a small piece", in exactly the same way that Montague grammar studied fragments. I would like to propose fragments with simple semantics, and then to study them as logical systems, doing all of the work of axiomatizing them, studying algorithms for all the computational aspects, etc. The requirement that the fragments be decidable means that one cannot go to first-order logic. Usually, we have to go for weaker logics. (But this is not always the case.) In my case, I'm harkening back to syllogistic logics and more up-to-date variants. So in a sense, a spinoff of natural logic is the re-examination of the syllogistic. But overall, I think that looking at small fragments of language could cause us to change our view of the relation of logic and language. For example: If one has a complete logical system for a fragment, then one might well take the logical system to *be* the semantics in some sense. (Whenever I raise this point in seminars, I get the most resistance on this from philosophers of logic, and so I believe that this opinion touches a nerve!)

And a next step in this area, a somewhat radical one, would be to go beyond the model-theoretic tradition in semantics. We also want to explore the possibility of having proof theory as the mathematical underpinning for semantics in the first place. This view is suggested in the literature on philosophy of language, and perhaps even philosophy of logic. There are some people who are doing this, notably Nissim Francez. Further, one can imagine serious treatments of linguistic phenomena like presupposition.

Incidentally, the logical systems in the area of natural logic are beginning to be implemented. We can look forward to computer systems which manage inferences, counter-model generation, and other things, all for reasonably interesting fragments of language. This suggests that the fragments themselves could be the vehicles to teach the basics of logic. The experience here is mixed, and I think that more will need to be done before we can take extended syllogistic logics as serious topics of the classroom.

From this vantage point, it is clear that the area of philosophy of logic is an active player in a number of topics of great human importance. It is closely related to philosophical logic, and hence is practically the same as the theoretical side of AI. My suggestion to the field is not to lose track of the original motivations of the field of logic: the study of reasoning and representation, of language and meaning.

15

Catarina Dutilh Novaes

Associate Professor
Faculty of Philosophy, University of Groningen

---

1. Why were you initially drawn to the philosophy of logic?

I'm afraid I'll have to elaborate a bit on auto-biographical details to answer this question, all the way back into my high school years. In high school, my favorite subjects were mathematics and literature (biology and history coming next). I was fortunate to complete two of my high-school years at a French Lycée in Paris, where both math and literature were taught at a fairly high level. What happened was that I was blown away by the beauty of mathematics; it was particularly striking to me how different methods and approaches would still converge into the same 'truths' (say, analytic geometry and algebra). So from very early on, I was intrigued by the magic of mathematics and the beauty of proving theorems. Still, people kept telling me that I couldn't pursue the two paths at the same time, mathematics and literature: I had to choose only one path.

It was almost by chance that I went on to study philosophy at university (in the meantime, back in Brazil). But as it turned out, philosophy allowed me not to have to make the choice everybody kept telling me was inevitable! The kind of philosophy I was taught as an undergraduate was heavily historical, à la française, with very little focus on analytic philosophy. But we did have the usual baby logic courses, and I also took quite a few courses at the math faculty (which is why I usually say I have a minor in mathematics, even though in Brazil we don't have the major/minor system). As an undergraduate, I was interested in philosophy of logic, philosophy of science and philosophy of language, the problem of induction in particular. But I was also already interested in the question of why formal methods and formalisms like mathematical formalisms can be applied to phenomena in general, and in fact deliver novel knowledge about the phenomena (in a sense, this is the very question that I attempted to answer in my 2012 book *Formal Languages in Logic*).

But when I graduated, it was clear to me that I did not know enough logic to be able to become a good philosopher of logic, so I enrolled in the Master's in Logic program at the ILLC in Amsterdam. Despite having some background in mathematics, joining this program was a huge shock: I had no idea of the extent of my ignorance! This was a difficult year for me, but also a very rich period where I learned an awful lot. For reasons still somewhat mysterious to me, I had the idea of choosing Ockham's theory of supposition as the topic for my Master's thesis: I had followed one course on medieval logic back in Brazil, but until then it had not been a strong research interest of mine. However, given my strength in the history of philosophy acquired back in São Paulo, the topic seemed to be my best shot at coming up with a decent thesis to satisfy the ILLC crowd. It turned out to be a decision with long-lasting consequences: I went on to write my PhD on medieval logic (which came out in book form as *Formalizing Medieval Logical Theories* in 2007, with Springer), and most of my published work is on medieval logic. A lot of it consisted in formalizations of medieval logical theories with modern logical tools, and the question of what warrants the 'fit' between formal and informal realms was always in the back of my mind.

However, although I was mostly working on historical topics, the questions driving my investigations were always predominantly systematic. The history of logic and philosophy seemed to me to offer a privileged vantage point from which to think about the fundamental philosophical questions pertaining to logic: What is logic? Why does logic work for us? How is logic a reliable tool for the production of knowledge? So while I continued to work on the history of logic, medieval logic in particular, I also wanted to pursue these systematic questions.

More recently, my research took what could be described as an 'empirical turn'. Almost by chance, I stumbled upon the literature on the psychology of reasoning, and realized that to answer the questions I was asking myself – at that point in particular, pertaining to the cognitive impact of reasoning with formal languages and formalisms as notational devices – I would just *have* to go empirical. This happened at the end of 2008, and since then attention to empirical approaches to cognition has become an important component of my research.

But I digress! To sum up, what drew me to the philosophy of logic was the experience of the almost eerie beauty of mathematics in high school, of the idea of proving theorems, and the power of formal approaches for the production of knowledge about phenomena in the world: in a word, logic as a *tool*. I wanted (and still want!) to understand how this comes about.

So I like to raise very basic questions about logic, reasoning and cognition. In a letter to his lover Lady Ottoline Morrell, Russell says the following about Wittgenstein: "He doesn't want to prove this or that, but to find out how things really are." (quoting from M. Potter's book *Wittgenstein's Notes on Logic*, p. 50) Now, I see myself as a person who doesn't want to "prove this or that" (I am not particularly good at proving things anyway). While I am a big fan of formal methods in general, I also believe that some of the most important philosophical issues pertaining to logic need to be investigated not by 'proving this or that', but by asking very fundamental questions – in fact, questions about the very practices of 'proving this or that'!

2. What are your main contributions to the philosophy of logic?

I suppose one of my main contributions to the philosophy of logic is the idea that different methodological vantage points are needed if we really want to make sense of a number of philosophical issues pertaining to logic. My work as a historian has taught me that concepts that are central to current developments in (philosophy of) logic all come from 'somewhere', i.e. are historical constructs, and understanding where they come from is crucial for a philosophical discussion of what they mean to us now. So in my current project, 'The Roots of Deduction', we are going back to the historical origins of the concept of deduction (understood as having necessary truth preservation as its cornerstone) in Greek mathematics and philosophy, in order to understand what motivated people to think that necessary truth preservation was an interesting property for an argument to have in the first place.

What empirical research on human reasoning has shown is that deduction and classical logic are not in any way accurate descriptive models of how humans reason, and, in particular, that the concepts of necessary truth preservation and indefeasible argument are something of a cognitive oddity (as also argued, for example, by Stenning and van Lambalgen in their 2008 book). So this suggests that there must be an interesting story to be told concerning the *historical* emergence of the deductive method. At this point, our working hypothesis is that this emergence is closely connected to the context of *debates* in ancient Greece.

Now, an investigation of the concept of deduction is a typical example of an issue which, to my mind, can only be adequately investigated with what I refer to as an integrative methodology besides attention to the history of logic and philosophy on the one hand, and empirical work on cognition on the other hand, another important component would be the more familiar formal methods, which are irreplaceable ways of generating important novel insights. Personally, I'm not doing much on this front at the moment, but it remains a crucial component for this kind of investigation. (I plan to look more carefully into some semantic frameworks developed for non-monotonic logics in order to conceptualize the divide between indefeasible and defeasible reasoning in terms of adversarial vs. cooperative dialogical games.)

To be sure, I am not the first one to emphasize the importance of the history of logic for the philosophical, systematic discussion of contemporary issues pertaining to logic. My main inspiration here has been Stephen Read. In fact, when I was in the second year of my PhD, I started reading his work, and it blew me away. I thought to myself: this is *exactly* what I want to do when I grow up! Since then, I've been fortunate enough to collaborate quite extensively with Stephen on a number of things, mostly but not exclusively on medieval logic. My former thesis supervisor in Leiden, Göran Sundholm, is also someone who emphasizes the importance of the history of logic, and from him I also inherited a certain proof-theoretical inclination (as well as the coolest academic genealogy ever: Veblen-Church-Turing-Gandy-Sundholm-me).

Something else that could perhaps be described as a contribution of mine is to emphasize the role of logical theories as *tools* for reasoning and intellectual inquiry – maybe something that could be described as a *usage-based approach*. Regarding medieval logical theories, for example, say supposition theory, when I first got interested in them, most of the literature had focused on investigating the properties of such theories from the point of view of modern logical theories, without raising the question of what medieval theories of supposition were theories *of* for the medieval authors themselves. Why did they need such theories in the first place? What use did they make of them? This led me to a different account of supposition theory as a theory of sentential meaning rather than as a theory of reference (as it had been described by modern commentators until then), which I develop in chapter 1 of my 2007 book and in a few articles.

With my current 'Roots of Deduction' project, we are adopting a similar approach. The question is: what practical/pragmatic circumstances were behind the emergence of the concept of a deductively valid argument? What was such an argument good for? At the beginning of the *Prior Analytics*, Aristotle says that a *syllogismos* (a term that is difficult to translate, as it is wider than what we now understand by 'syllogism', but is not exactly the same as our current notion of 'deduction') is a discourse (*logos*) in which, certain things being stated, other things follow *of necessity* from their being so. It is tempting to think that the 'of necessity' clause corresponds to our modern notion of necessary truth preservation, but this seems to be mistaken. For example, Aristotle clearly does not accept reflexivity as a property of these arguments, so 'A implies A' is not a good *syllogismos* according to him. As mentioned above, my current hypothesis is that one must think of such arguments as useful tools in dialectical contexts of debates and disputation, so that the 'of necessity' clause suggests an argument which compels the interlocutor(s) to accept (the truth of) the conclusion, if they have accepted (the truth of) the premises – something that, more than 2000 years later, was described by Wittgenstein as the 'hardness of the logical must'.

Similarly, in my research on formal languages which culminated in my book *Formal Languages in Logic* (2012), the question I asked myself was: what kind of cognitive impact does the use of formal languages and other formalisms have for reasoning? Again, I wanted to focus on formal languages not only as mathematical objects, but also as tools for reasoning. This led me to engage with the literature on extended cognition from philosophy of mind, and of course with a wealth of empirical results on human reasoning and cognition from psychology and cognitive science.

The 20th century is marked by a focus on logical systems and logical theories as things that we reason *about* (the meta-theoretical turn with Hilbert et alia); we seem to have neglected the dimension of logical systems as things to reason *with* (this is also something I've learned from my former supervisor Sundholm). My emphasis on *uses* of logical theories is loosely inspired by the later Wittgenstein (whereas the Wittgenstein of the *Tractatus* represents precisely the radical expulsion of the thinking subject from the realm of logic!), but ultimately it is the question that I've been asking myself all these years: how can logical frameworks be such powerful cognitive tools for the production of new knowledge about real phenomena?

I don't expect my usage-based perspective to overturn the well-entrenched meta-theoretical paradigm, but if I manage to convince at least some people to think about logic also from the point of view of the uses we make of it, this will have been a small contribution. In fact, I certainly do not aim at a complete extermination of the meta-logical paradigm; it has delivered and continues to deliver astonishing, fascinating results – Gödel's theorems being perhaps the best example. In fact, some of the most interesting results in the meta-logical tradition pertain precisely to an investigation of the *limits* of the formal methodology as such.

In terms of methodology, I'm a bit of a Feyerabend kind of person: the more, the merrier. Hence, my goal is to add the usage-based perspective to the pool of respected approaches in the philosophy of logic, not to defend the view that it is the only legitimate approach.

3. What is the proper role of philosophy of logic in relation to other disciplines, and to other branches of philosophy?

This depends on one's conception of logic in the first place. If what counts as logic is viewed exclusively as the highly technical work that has been done in mathematical logic since the early 20th century (say, the sort of thing that gets published in the *Journal of Symbolic Logic*), then most of it will have only tenuous direct philosophical significance – which is not to say that this is a bad thing. This research tradition has to some extent been emancipated from its philosophical origins by becoming more and more mathematical, and this is simply how things go; things change.

But there is also exciting work being done with the methodology of logic on philosophically important questions such as the notion of truth, much of what is done within formal epistemology, etc. Generally, I believe that a large number of philosophical questions can be fruitfully investigated with logical frameworks. Some examples are Kripke's groundbreaking work in modal logic, more recently work on axiomatic theories of truth (also inspired by Kripke's later work), formal modeling in ethics, epistemology, metaphysics, etc. In all these cases, the general question of how formalisms can indeed be used to investigate these extra-logical phenomena – what warrants the application of logical tools to these issues – must be addressed, and this is a question for the philosopher of logic par excellence.

Moreover, if one thinks (as I do) that the philosophy of logic is the right place to discuss a wide range of phenomena pertaining to human reasoning in general, then philosophy of logic should be seen as relevant for pretty much all other disciplines of scientific inquiry. (But of course, this is precisely the kind of statement a philosopher of logic is expected to make.)

More precisely, some of the most exciting work in logic at the moment takes place at its intersections with other disciplines. Since the 19th century, logic has enjoyed a privileged relationship with mathematics, but now computer science and linguistics are perhaps even more relevant areas of interaction for the logician. At any rate, and in the spirit of my defense of integrative methodologies, to my mind interdisciplinary collaboration is very much to be encouraged. But for researchers in other disciplines to interact fruitfully with philosophers of logic, it would be important for philosophers of logic themselves to be more open-minded concerning what counts as questions belonging to the remit of their area. I've often heard from philosophers that the kind of work I do on the interface with psychology is not 'philosophy', properly speaking. Such exclusionary attitudes are not likely to win us many friends out there.

4. What have been the most significant advances in the philosophy of logic?

It is difficult and possibly not very fruitful to talk about advances in the philosophy of logic in isolation from advances in logic proper. Going back a bit in time, I'd say that one of the greatest advancements in logic of the last many centuries was the invention of formal languages and logical formalisms. To be sure, this was a long and complicated gestation, which I discuss in chapter 3 of my 2012 book. But it fundamentally changed the way we do logic, when compared to traditions which did not make extensive use of special notations such as the ancient Greek and Medieval Latin traditions.

More recently, and more specifically in the philosophy of logic, one of the most important events has been the revival of debates on logical consequence provoked by Etchemendy's book *The Concept of Logical Consequence* (1990). He raised some very important questions, and even though his answers are not always satisfying, the book certainly forced us to go back to some of the most basic philosophical questions pertaining to logic and think harder about them. Some of the most significant papers in philosophy of logic of the last 20 years have been responses to Etchemendy's challenge.

Another important development has been the emergence of paraconsistent logics (again, precisely the kind of thing you'd expect to hear from a Brazilian philosopher of logic!). Since the birth of logic with Aristotle, consistency and contradiction have been largely viewed as the building blocks of logic. But paraconsistent systems show that it is possible to be inconsistent at times without being entirely 'illogical', in particular by avoiding triviality. From a different angle, people such as Branden Fitelson are now talking about how coherence and consistency may not be reasonable norms for belief after all (in light of e.g. the preface paradox), so there is a whole ongoing debate on what consistency really means and what it is good for. This is a favorable turn of events, as logicians have tended to be quite dogmatically obsessed with consistency. (Recall David Lewis' famous claim to the effect that, once one gives up consistency, meaningful debate is no longer possible.)

I also think that formal/axiomatic theories of truth have made important philosophical contributions. One of them is the observation that adding the truth predicate to a given theory (usually, the toy example is Peano Arithmetic) yields conservative extensions of the theory: there is nothing involving the 'old' vocabulary of the theory that can be proved in the extended theory which cannot be proved in the original theory. This suggests that the truth predicate does not add anything substantive in terms of provability. On the other hand, it has also been observed that adding the truth predicate generates a 'speed-up' effect: one can formulate much shorter proofs of a number of theorems in the extended theory than in the original theory. So this is somewhat in tension with the conservativity results: the truth predicate does seem to add something substantive after all in terms of expressive power, even if not in terms of absolute provability. The debate is still ongoing, and much of this work is very technical and somewhat baroque: at times, the connection with the philosophical notion of true is somewhat lost. Still, to my mind, the conservativity and speed-up results have provided important philosophical insights on the notion of truth.

Another burgeoning field of inquiry within logic is the emergence of multi-agent logical systems, such as in the work of e.g. Johan van Benthem and collaborators. In the last couple of centuries, the focus in logic has been – tacitly or explicitly – the endeavors of the lone agent. But arguably, at the beginning logic was essentially a multi-agent enterprise, pertaining to practices of argumentation and debate in ancient Greece. It also had a strong multi-agent, dialogical component in the Latin Middle Ages, in particular with different forms of oral disputations such as *obligationes*. Now, with these new multi-agent systems, the range of application for logic suddenly becomes much larger, as it now encompasses social phenomena and rational interaction. More traditionally minded philosophers of logic have not yet picked up on the philosophical relevance of this switch from mono- to multi-agency, but it is to be hoped that it is just a matter of time. (Notice that mono-agency can be viewed as a limit case of multi-agency, but the converse is much less straightforward. This suggests that the multi-agent perspective should be seen as the more general framework.)

5. What are the most important open problems in philosophy of logic, and what are the prospects for progress?

There has been much attention in recent literature to paradoxes. The Liar paradox remains the king among the paradoxes, but there are a number of other paradoxes, Curry's paradox in particular, that have received quite some attention recently. Still, we do not seem to have attained a deeper philosophical understanding of what paradoxes really mean for logical theorizing. Paradoxes are typically rather extreme cases, which are not likely to crop up in 'ordinary' occurrences of reasoning and argumentation: who ever says, 'What I am saying now is false', in ordinary contexts? But proposals to block the occurrence of paradoxes usually entail a high level of revision, and often we end up with logics which are no longer obvious tools for reasoning or arguing (say, Field's system in *Saving Truth from Paradox*).

So I think we still need to attain a better understanding of why it is (*if* it is) imperative to block the emergence of paradoxes, and how far one should be willing to go for this end. In a sense, paradoxes are like a cold, which somewhat hinders the functioning of the organism but does not prevent it from functioning well enough. Some solutions to paradoxes, however, are like very invasive therapies, say treating a cold with chemotherapy or an amputation, and arguably do more damage than the original ailment they were supposed to treat. In this respect, the work of Graham Priest is a good illustration of someone who is not afraid of paradoxes and truly attempts to understand their nature rather than coming up with quick fix-ups. In fact, his formulation of the Inclosure Schema as the general structure underlying paradoxes (in *Beyond the Limits of Thought* and a few articles) certainly counts as one of the most important advancements in philosophy of logic of the last decades.

Another important and deeply philosophical question is: what are logical systems models *of*? In his book, Etchemendy makes a big deal out of the idea that the Tarskian account of logical consequence does not capture the 'intuitive notion' of logical consequence, but without telling us exactly what this intuitive notion really is. (You might think: since it is intuitive, it requires no explanation. But intuitive for whom? Does everyone in fact share it?) This has been pointed out by e.g. Stewart Shapiro, who then asked the crucial question of what logic is about, what it is a model *of*. (I often say to Stewart that he asks all the right questions – to which he replies 'But you think I give all the wrong answers!' Well, not exactly...)

At the same time, the empirical results I've mentioned above show that human reasoners do not really reason following the canons of traditional logic, so the idea that logic provides a descriptive account of human reasoning is simply untenable. Some say that logic provides a *prescriptive* account for human reasoning, but this view is also deeply mistaken in my opinion (and in the opinion of G. Harman and a few others). So it is quite surprising to realize that in a sense we do not really know what logic is about! Ok, I am exaggerating a bit here, but I'd like to see much more discussion on the so-called 'intuitive notions' (of logical consequence, of validity etc.), which are then purportedly captured by logical formalisms.

It will not surprise anyone if I say that, in my opinion, the most promising approach to make progress on this issue is to go back to the history of logic and philosophy and isolate the historical sources of these so-called 'intuitions'. So for example, my entry on medieval theories of consequence at the *Stanford Encyclopedia of Philosophy* takes as a starting point Tarski's condition (F) of adequacy for theories of logical consequence, and attempts to retrace some of the theoretical and historical origins of what Tarski described as the 'common' notion of logical consequence in these Latin medieval theories. By engaging in what I call an exercise of conceptual genealogy, it becomes clear that these so-called intuitive notions often have long and complex histories, and are not 'pre-theoretical' properly speaking: usually, a significant amount of theorizing is involved in these developments.

I would also like to see more collaboration between cognitive scientists and psychologists, on the one hand, and philosophers of logic and logicians on the other hand, for the empirical investigation of human reasoning. Everyone has much to gain from such a collaboration, as illustrated for example by the Stenning & van Lambalgen collaboration (e.g. their fantastic 2008 book with the unremarkable title *Human Reasoning and Cognitive Science*). Generally speaking, I am an advocate of empirically informed approaches in philosophy across the board, and for the philosopher of logic, the obvious partner seems to be the cognitive scientist studying human reasoning.

All in all, I would say that there is much work to be done to keep us busy in the coming years and decades. I would advocate for a broader range of topics to be addressed and for a broader range of methodologies to be used in philosophy of logic.

16
Ahti-Veikko Pietarinen

Professor
University of Helsinki, Finland

The following questions are answered succesively:

    Why were you initially drawn to the philosophy of logic?

    What are your main contributions to the philosophy of logic?

    What is the proper role of philosophy of logic in relation to other disciplines, and to other branches of philosophy?

    What have been the most significant advances in the philosophy of logic?

    What are the most important open problems in philosophy of logic, and what are the prospects for progress?

I grew up in very philosophical surroundings and hence it was not too surprising that, whatever the subject of research was going to be for a beginning graduate student, it would involve a strongly philosophical element. My formal training had been in the exact sciences, so logic and philosophy of logic made their way into my research without much further thought.

A fair amount of today's work in logic is not relevant to philosophical concerns, though some of it may turn out to be relevant to philosophy in the future. Making predictions is like Barbara Hutton's marriage plans. But in general, there has been a steady drift of logic away from philosophy and philosophically motivated insights. G. H. von Wright wrote about this in 1994 in "Logic and Philosophy in the Twentieth Century", and he advised me to continue work on "philosophical logic". Hintikka's works in logical philosophy, largely unparalleled in their breadth and depth, had already begun to stir my curiosity that time.

There is nothing deplorable as such in logic's unphilosophical trend. Logic is one of those (perhaps the only one) truly transferable skills that can really aid progress in various parts of the sciences, whether they concern the nature of computation, cognition or communication, physical processes, biological processes, or the origin and evolution of cultures, meanings, languages and behaviours. Yet it is quite salient how sightless, and I mean philosophically and historically sightless, much of the recent work in logic has been, even in such areas as "philosophical logic". This is where the philosophy of logic enters the scene.

My own research has been guided by the belief that one should check and learn everything that has been done on the topic of investigation first, before moving on. Often this has meant digging deep in the archives of unpublished papers. Mostly I have been concerned with 19th and early 20th century logicians, especially Charles Peirce, and his remarkable students and contemporaries. My goal has been to become better acquainted with the content and context of those early discussions. The archives are still loaded with surprises. Every so often, these exercises have confirmed the wisdom that, for any idea presented, say during the last 100 years, there is another one anticipating it. This happened to me with respect to the invention of first-order logic, the ideas of soundness and semantic completeness, the theory and semantics of quantifiers, the origins of systems of modal logic, quantified modal and epistemic logic, the question of expressing mathematics in higher-order logic, the semantic and pragmatic theories of relevance, and conventional and conversational implicature. All of these had their pre-1913 ancestors [2005b, 2006a, 2014d]. No conspiracy here: it is just that the canon is too viscous to adequately reflect the real state of affairs. Sometimes, we may need the distance of 100 years to grasp what was going on way back when.

For my dissertation [2001] I worked on *independence-friendly* (IF) logics, which Hintikka had started to develop in the 1970s. I identified semantic structures for various types of information flow in IF formulas by means of *game-theoretic semantics* of imperfect information and beyond. I had noted, in my 1997 MSc thesis which was on IF epistemic logics in knowledge representation, that the pertinent semantic games are those of imperfect recall, and exhibit forgetting of actions and the forgetting of information [2001b]. From such considerations some generalizations follow, such as 'intensional' informational independence, which conditionalizes the existence of imperfect information or recall on particular actions that the players take, and also conditionalizes it on learning, which is an increase in the information of the teams (consisting of 'multi-selves' or individual members). Moreover, hiding negation via in*complete* information games, or congruent models for IF modal logic, or non-standard semantic games for different kinds of partial and paraconsistent logics, presented themselves as promising new ideas [2000, 2001c, 2004a, 2005a]. Some of those ideas were rediscovered and pursued by others. An anecdote here: when I presented the ideas of IF logic to my MSc supervisor at the University of Turku

– this was in 1996 – he immediately replied: "Why don't you go on to denote in the language the *dependence* of quantified variables instead of independence? That seems more natural." As I further mused to him that IF logic cannot be compositional, he immediately not only denied that but sketched the context-carrying method of turning it into a compositional language. As the knowledgeable reader here will note, these ideas appeared in the market as dependence logics, compositional semantics, and so on.

Informationally independent (IF) extensions of first-order epistemic logics have turned out to be philosophically crucial to understanding the nature of the objects of our knowledge. I showed in [2001a] how they solve the problem of intentional identity, that is, they analyse the meaning of Geach's Hob-Nob sentences and their variants by new models of quantified epistemic IF logic. One can further establish a match between quantifier independence on epistemic modalities and imperfect information in the associated semantic games [2010b]. This requires a new kind of uniform-domains assumption. But the conceptual side of the situation is more interesting. Trans-world identities are the results of player-level ignorance. Hence rigid designation turns out to be a derivative and degenerate form of those world-line identities that are constitutive of objects of agent-level specific knowledge. That is, logical, semantic and linguistic notions fall out of the pragmatics of action.

Because 2013 is the year of the Grice Centennial, let me also mention the following little-known fact. In [2004b] I put forward a hypothesis that Grice's work was influenced by Peirce's theory of signs and communication and his pragmatistic logic and philosophy. The evidence concerned several cases in which Grice's explanations for the central notions of his theory of meaning and logic of conversation had close correlates in Peirce's theory. For example, Grice's 1967 conversational implicatures and cooperative principle are just like Peirce's 1905 characterisations of speaker meaning. On Peirce's 1905 view, speaker meanings consist in the intention to fix the implications and non-implications of assertions together with the belief that the utterer may have succeeded in doing so. Peirce and Grice both agreed that the utterers and interpreters have a common rational purpose and that the interpreter is expected to recognize that the utterer is present both in the utterance and as a deliverer of it. They also agreed that working out particular conversational implicatures requires a common ground constituted by common knowledge. Grice even talks about iconic, associative and conventional modes of correlation, which is essentially Peirce's icon-index-symbol trichotomy. Grice in his published work never referred to Peirce. But my hypothesis was recently conclusively established when the Grice archives turned up a set of lecture notes for a course Grice gave at Oxford in the early 1950s. The title of those lectures was – "Peirce's General Theory of Signs" [2013d]. Grice actually had translated Peirce's terms into his own lingo: Peirce's "signs" became Grice's "meaning", Peirce's "interpretants" "implicatures" and so on! Grice's first sketches of the definition of speaker's meaning in terms of intentions appear towards the end of the lectures. So what became the springboard for contemporary pragmatics was itself rooted in a reworking of Peirce's theory of signs, pragmatistic meaning and signification.

It is also worth noting that a characterisation of common knowledge as an infinite series of mutual familiar interactions between utterer and interpreter is present in Peirce's unpublished manuscripts from 1908. That notion is of course crucial in Grice's logic of conversation. Later on Lewis, Aumann, FHMV and scores of others picked up the idea that became important in contemporary logic, game theory, and philosophy of language.

Here an open question arises. Can game theory really provide a general theory of rational communicative interaction? The answer in the general case is not known. Which of our linguistic and non-linguistic competencies in communication can be modeled as aspects of strategic reasoning? If communicative action is not just a move but a belief, game theories need a radical refitting if they are to serve as theories of choice accommodating mental, cognitive and semantic entities – for the needed games either lack solution concepts or do not conform well to social contexts of linguistic interactions. Just showing how Grice's maxims emerge from endogenous considerations of meaning does not get us to the bottom of the issue. And merely treating belief changes as variants of Bayesian update protocols on incomplete information is equally ineffectual. Rather, it seems to me that the key to how to encode intentions into strategic interactions is of the same nature as how to model *abductive* forms of reasoning, how new conceptual categories are created, and how to cope with *ex ante* fundamental uncertainties when the problem contexts are neither well-structured nor invariant. These are very different issues from what have traditionally been perceived as species of inductive problems.

One of the reasons why logic has lost its philosophical status has to do with two disparate conceptions of logic, only one of which was deemed worthy of serious theorising. The distinction goes back to the Arab and medieval logicians, even to Aristotle, though it was largely forgotten in the hubris of the last century's formalistic and symbolic race. Cognitive neurosciences, implicit information processing, unawareness, etc. have partially revived this old idea of the possibility of theorizing about the nature of the instinctive faculty of logic, the *logica utens*, as opposed to the educable, theoretical logic, the *logical docens*. I have worked on an agenda for the recovery of what the *logical utens* could be, arguing that in that faculty we may find the relevant components of the logic of discovery and its abductive (retroductive, adductive) modes of reasoning [2005c]. The imaginative, iconic and metaphoric modes of logical thought seem to be the missing parts of the puzzle.

Visual and non-symbolic representational systems are becoming increasingly important in logic as well as in the related fields of computing and cognitive science. Peirce had proposed logics for representing and reasoning about "actions of the mind in thought" using specific kinds of iconic signs which gave rise to the diagrammatic logics – the *Existential Graphs* (EGs). They were studied by C.K. Ogden and F.P. Ramsey in England, but were soon forgotten due to a cascade of alternative impulses. During the last ten years or so, I have explored the key notions of these diagrammatic systems, many of which were never published by Peirce or by those who later on could have done so [2014c].

Specifically, I have endeavoured to bring out a number of hitherto unknown historical and logical facts concerning both the genesis and the reception of the method of EGs [2011b, 2014b]. Peirce conceived these diagrammatic logics in 1896, but even today their investigation has barely begun. By 1903, Peirce had contemplated possible-worlds and game-theoretic semantics as interpretations of his new systems of modal logic and rejected the accessibility relation in favour of a continuity interpretation inspired by his explorations in early topology. Moreover, the logic of quantified modalities, as developed in the gamma part of the EGs, lies behind Peirce's restatement of the principle of pragmaticism ("The possible is what *can* become actual"). As Peirce saw it, the EGs present us a method for the analysis of reasoning not exhausted by what is expressed by any particular formalism or a visual system of diagrams. Graphs represent *imaginary* diagrams of relational facts aided by conventions. You can scribe both the premisses and the conclusion on the one space of assertions and sit back and enjoy the "moving picture of thought" that displays proofs as continuous illative transformations. To take these graphs to correspond or be translatable to the known systems of symbolic logic is to oversimplify the structures that iconic representations are able to capture. For example, the lines of identity have to do not only with the existence, identity, predication and co-reference, but also with *identification* (when viewing the films starring quantified modal logics), and substantive and qualitative possibilities (when viewing the films starring higher-order logics). Beyond the proof-theoretic entertainment, these logics can and were used by Peirce extensively in his analyses of some core issues in the philosophy of logic and language: the logical interpretation of modalities, complex donkey-anaphora and Geach-Kaplan-Karttunen types of anaphora, mathematical argumentation, etc.

Quantified modal logics, then, have been around us much longer than we thought, as indeed have the allied philosophical problems that transworld identification gives rise to. But new issues also emerge. In EGs, quantifiers are denoted by the extremities of the lines of identity and they range over individuals that are in principle identifiable. In addition to quantification, these continuous lines serve as identifying devices. Of course, the failure of the Frege-Russell thesis – that the verb for *being* is three-way ambiguous among existence, identity and predication – is here imminent [2006b]. But there are further implications. Qualities that are not occurrences, that is, that do not exist and have no manifest duration, give rise to assertions that can bear non-propositional content. Thus, with the gamma modal logic we are really quantifying over the 'might-bes'; hence, feasible modal logics need to take such possible objects as constitutive of their real domains. Peirce's original idea of interspersing quantification with various modalities ("tinctures"), including epistemic modalities, was equipped with a 'model-theoretic' semantics that assigns philosophically robust significations to the diagrams of these logics. Since the 1950s, that problem attracted renewed interest in the debate about "quantifying into the scopes of modal operators." The irony is that symbolic representations fail to capture what Peirce's diagrams do.

In order to attain a comprehensive logic of icons, though, Peirce's diagrammatic logics need to be expanded in certain ways. This realization has fuelled my investigations on the logic of *images* and the logic of *metaphors* [2012a, 2013a]. Diagrams, the many-dimensional expressions that analyse meaning by virtue of the iconicity of logical form, do not alone suffice. We need metaphors that combine complex diagrammatic representations with images that serve as elementary constituents of qualitative spaces. The interpretation of images corresponds to the interpretation of non-logical vocabularies. This raises the question of whether images are also linguistic – in other words whether the simple qualities images partake of are the simple qualities of some propositional content. A real picture theory of language is thus one in which images interpret elementary characters of objects that constitute propositions.

The other expansion that I have presented is a theory of metaphors that takes metaphoric meaning to be a matter of iconic forms of logic. Like icons in general, metaphors evoke similarity considerations and representations of parallelisms in another media. The existence of metaphors is, in fact, a strong argument for the fundamentally iconic modes of logic operative in cognition and reasoning. Such modes stand in stark contrast to a number of prevailing theories in the philosophy of mind, language and cognition, including the language of thought hypothesis, meaning holism, and mental models and conceptual metaphor theories.

Overall, the ordinary means of conceptualising the notion of propositional content seems to me to be hopelessly outmoded and limits analytic philosophy of language's progress.

For these reasons, something like Neurath's vision for graphical communication through his ISOTYPE system is extremely interesting from the points of view of philosophy of language and logic. What were the key presuppositions of his picture language? I have argued [2011c] that Neurath was in fact a calculist about meaning and that his vision for such multi-modal modes of communication suggests that the Unity of Science programme might have surpassed the logical empiricists' doomed attempts – if only it had gone in the direction of the logic of images, a path that Neurath did not take.

It is, in fact, crucial to ask why logic was thought to be based on symbolic notations in the first place. In [2010a], I proposed logics that have no visual and no written appearance whatsoever: no symbols, no marks, no language. This is conceivable as soon as it is realised that diagrams constitute a category that includes more than systems of signs that are visually or spatially perceivable. One can, without too much effort, develop a propositional logic based on auditory signals instead of symbols or spatial diagrams. The only puzzling notion here is the negation – the absence of a sound – but so was the notion of the cut in valent graphs of chemical compounds a century ago.

What else may be hampering progress in the philosophy of logic? One could ask, for example, what is the rationale for scores of philosophers of language being obsessed with such singular ideas as the 'new theory of reference'? One could again introduce someone like Peirce, whose theory of proper names spells out both a non-descriptive theory of the denotation of proper names (which is wider than that provided by causal-historical theories), *and* a non-substitutional theory of quantification [2010c]. And Peirce's theory of names is not even his full theory of meaning. That theory, *pragmaticism*, is a consequence of several fundamental features of logic and the theory of signs. Pragmaticism holds that meaning consists in the meaningful entity's contribution to an expectation about future experience. A sound argument for the correctness of the pragmatistic theory of meaning can be worked out, building on several fundamental logical and semiotic assumptions: drawing on unpublished manuscript material, I worked out a reconstruction of Peirce's 'proof of pragmaticism' in [2005d, 2011a] that used contemporary game-theoretic notions.

Peirce defended the view, taken up by Ramsey among others, that logic is a normative science. What does this mean? Well, rules govern self-controlled action, all communication is in signs, logical thought and habits are self-controlled, and logic is semeiotic. Self-controlled agents have normative ideals in their minds when they engage in conversation about the meanings of intellectual signs and purports. Now these mysterious 'habits' are many-world entities that link situations to actions, also at the off-equilibrium, while the rules of meaning-constitutive practices and activities provide the logical and strategic structure for them. (Normativity not grounded in such rules governing the meaning-constitutive practices is in fact inconsistent [2012b].) This argument lies at the core of the proof of pragmaticism. Moreover, the normativity of logic also invites us to reconsider the rationale for logics that aim at capturing or modeling the action, behaviour and human reasoning in the real world. Is it not the responsibility of logic to pursue the ideals rather than to fit models into the Procrustean bed measured by insignificant details about the complexities and insecurities of the actual world, and the singular beliefs and idiosyncrasies of concrete players?

Such (Peircean) conceptions of logic and pragmaticism have turned out to be virtually those of Hintikka's game-theoretic semantics – including the idea of players "feigned in our make-believe" [2003]. Investigations along these lines focus on logic insofar as it capable of providing analyses of normative, conventional, habitual and strategic action. (As previously mentioned, Grice erected his theory of conversation on a Peircean foundation.) But what is important to note here is that cooperation is a property of model-building games and an integral part of Peirce's method. Cooperative model-building employs the same theoretical construct as competitive semantic games do. The two kinds of games, the semantic and the model-construction games, are, in fact, two sides of the same conceptual coin. For all sentences of the underlying language there exists a winning strategy for the utterer in the model-construction game if and only if there exists a winning strategy for him in the semantic game in a model [2013c].

In fact, general principles governing scientific and, specifically, mathematical practices are related to such model-building activities [2010d]. Let me mention one example. Sceptical currents in the philosophy of mathematics, like fictionalism, figuralism and other phantasms of the living fail to analyse the semantics of the specific tropes on which their arguments rest. This objection can be buttressed by mathematical pragmaticism. In its modalisation of mathematical entities substantive fictionalism assumes a metaphysical notion of modality that is not viable from the points of view of scholastic realism and fallibilism about mathematics [2013b]. Brouwer's languageless conception of mathematics raises another challenge. Pragmatistic philosophy of mathematics is also incompatible with mathematical structuralism, although both agree that it is the forms of relations that are of primary interest in the study of mathematical activities. Now it *is* the nature of these forms that logic studies, but logic cannot be stripped of its meaning, left as a pure mathematical form lacking significant constituents. The modalization of mathematical entities requires an ontology which includes real possibilities. So here pragmaticism as it has to do with experimentation and observation concerning the future expectations we have on the forms of relations (conceived as diagrammatic and iconic representations of mathematical entities) appears superior to mathematical structuralism. The pragmatistic approach to the philosophy of languages presupposes no mathematical foundations although it has such representations as its objects of study. These objects, substantive and qualitative possibilities, quantified by the continuous lines in the higher-order graphs, have a reality which fictionalism and structuralism both have denied [2014a]. (More generally, "there is no entity without qualitative possibility.")

Why have these important Peircean ideas gone largely unnoticed? One reason is that they are, even after Russell's own conflations a century ago, still associated with pragmatism. Pragmatism has a very uneasy relationship with logical theories. Peircean pragmaticism, on the other hand, *is* logic, conceived as semeiotic. In contemporary discussion, neglecting the logical roots of pragmatistic philosophy is yet another symptom of logic as language, as maximally universal medium of expression. Yet the two presuppositions concerning the role of logic in pragmatism – universalism and calculism – delineate two opposing philosophies, pragmatism and pragmaticism [2008]. Contrary to what is commonly assumed in the literature, it is in fact pragmaticism that is methodologically the more tolerant of the two, since pragmaticism permits the investigation of domains in piecemeal. Pragmatism, by contrast, arbitrarily and in an empiricist manner, aims at a classification of things into the logical and the extra-logical. Hence, pragmaticism embraces pluralism, while pragmatism does not.

This puts someone like John Dewey (a pragmatist but not a pragmaticist) and his amorphous logic *cum* metaphysics into a new perspective. Dewey's standpoint has been elusive, as his thinking apparently pulls in two opposing directions: his meta-systematic perspective on scientific inquiry fails to reconcile with his one-world presupposition about meaning. Ultimately, then, the tragedy is that Dewey missed the promising opportunity to evaluate problematic situations in terms of unactualized possibilities [2011d]. In the words of Benjamin Button: "Our lives are defined by opportunities. Even the ones we miss." Moreover, a one-world philosopher is by necessity a revisionist metaphysician whose image of her guess at the universe must satisfy some predetermined criterion. With high plausibility, such a vision is false, and leads to universalism and conceptual relativism. The alternative is the many-worlds philosophy, which permits indefinite variability of meaning and hypotheses, not unlike what used to be called descriptive, logic-approving metaphysics.

Not only may the diagrammatic, topological, homological, auditory, etc. logics be scientific assets; they are also promising from the educational point of view. Philosophy of logic may never be the first choice for every student's thesis work. But it also need not be confined to a formal-analytical project which all too often transmogrifies philosophical problems into exegetic forms of rational reconstruction. That kind of reasoning shackled humankind to the past. Sadly, in our age it is not the sham-free and fake-free reasoning that is held in high honour. The systematic and historical comparisons of pragmatistic philosophy with contemporary logic accentuates the continuum of logical ideas with the *Trivium*. It is not logic that has driven the philosophies of mind, language, mathematics and science into separate compartments of study. It is not logic that claims syntactics, semantics and pragmatics as distinct fields. In fact, the conceptual interplay between semantic and pragmatic aspects of meaning turns up a negative result when viewed from the game-theoretic point of view: that what is semantic and what pragmatic cannot be distinguished by rule-governed means [2007]. (The difference lies in the latter entertaining epistemic relationships with respect to the solution concepts and strategy profiles in the analysis of meaning.)

Why not introduce students to a modern philosophical *Trivium*, logic in its broad sense, and make it a compulsory module of training in liberal education? This means not just promoting what the euphemism of critical thinking tries to camouflage, for as it is usually practiced and taught critical thinking surely leaves much to be desired. In my view, critical thinking can come to fruition only if proper attention is paid to the discovery of arguments, rather than to the evaluation or justification of arguments after the fact. It is the discovery of arguments that ought to be the first goal of reasoning. And critical thinking belongs to logic conceived in the wide sense. Maybe logic is the methodology for all forms of reasoning, human and non-human. Logic understood in this wide sense aims at a meta-theory for sciences, and strives to unravel the mysteries of scientific and artistic innovation, ampliative forms of reasoning, and interactive and communicative behaviour.

We might thus do well to resist von Wright's pessimistic remarks and take logic in its wide sense, not as an exercise in different kinds of formalised languages with unanalysed semantic components, and not as a tool for distinguishing correct forms of human reasoning from incorrect ones, but as that business devoted to the search for the principles on which the most general forms of timeless thought depend.

Bibliography

2014d. "Peirce's Systems of the Gamma Modalities", *International Review of Pragmatics*.

2014c. (ed.) *Logic of the Future: Peirce's Writings on Existential Graphs*.

2014b. (ed.) *Synthese*, special issue on *Peirce's Philosophy of Logic and Language*, Pietarinen, A.-V. (ed.).

2014a. "A Scholastic-Realist Modal-Structuralism", *Philosophia Scientiae*.

2013d. "Grice's Lecture Notes on Peirce's Theory of Signs", *International Review of Pragmatics*.

2013c. "Logical and Linguistic Games from Peirce to Grice to Hintikka", *Teorema*.

2013b. "Two Challenges for Fictionalism in Mathematics". *Journal for General Philosophy of Science*, in press.

2013a. "Iconic Logic of Metaphors", *Journal of Cognitive Science*, in press.

2012b. "Why Is the Normativity of Logic Based on Rules?", in C. De Waal & K.P. Skowronski (eds.), *The Normative Thought of Charles S. Peirce*, Fordham: Fordham University Press, 172-184.

2012a. "Peirce and the Logic of Image", *Semiotica* 192, 151-161.

2011d. "A Hedgehog Who Thought He Was a Fox: Dewey betwixt the One and Many-World Philosophies", in L. A. Hickman, M. C. Flamm, K. P. Skowronski & J. A. Rea (eds.), *The Continuing Relevance of John Dewey: Reflections on Aesthetics, Morality, Science, and Society*, Amsterdam: Rodopi, 225-241.

2011c. "Principles and Practices of Neurath's Picture Language", in O. Pombo, S. Rahman & J. M. Torres (eds.). *Otto Neurath and the Unity of Science* (Logic, Epistemology, and the Unity of Science 18), Dordrecht: Springer, 71-82.

2011b. "Existential Graphs: What the Diagrammatic Logic of Cognition Might Look Like", *History and Philosophy of Logic* 32(3), 265-281. (Translated in Chinese, *Philosophical Analysis*.)

2011a. "Moving Pictures of Thought II: Graphs, Games, and Pragmaticism's Proof", *Semiotica* 186, 315-331. (Translated in Portuguese.)

2010d. "Which Philosophy of Mathematics is Pragmaticism?", in M. Moore (ed.), *New Essays on Peirce's Mathematical Philosophy*, Chicago: Open Court, 59-80.

2010c. "Peirce's Pragmatic Theory of Proper Names", *Transactions of the Charles S. Peirce Society* 46, 341-363.

2010b. "How to Analyse Rigid Designation? World Lines and Imperfect Information", in Arrazola, X. & M. Ponte (eds.), *Proc. Second ILCLI International Workshop on Logic and Philosophy of Knowledge, Communication and Action*, San Sebastian: University of the Basque Country Press, 351-369.

2010a. "Is Non-visual Diagrammatic Logic Possible?" In A. Gerner (ed.), *Diagrammatology and Diagram Praxis*, London: College Publications.

2008. "The Place of Logic in Pragmatism", *Cognitio* 9(1), 247-260.

2007. "The Semantics/Pragmatics Distinction from the Game-Theoretic Point of View", in A.-V. Pietarinen (ed.), *Game Theory and Linguistic Meaning*, Oxford: Elsevier Science, 229-242.

2006b. *Signs of Logic: Peircean Themes on the Philosophy of Language, Games, and Communication*, (Synthese Library 329), Dordrecht: Springer.

2006a. "Peirce's Contributions to Possible-worlds Semantics", *Studia Logica* 82, 345-369.

2005d. (with L. Snellman) "On Peirce's Late Proof of Pragmaticism", in T. Aho and A.-V. Pietarinen (eds), *Truth and Games*, Helsinki: Acta Philosophica Fennica 78, 275-288.

2005c. "Cultivating Habits of Reason: Peirce and the *Logica Utens* versus *Logica Docens* Distinction", *History of Philosophy Quarterly* 22, 357-372.

2005b. "Compositionality, Relevance and Peirce's Logic of Existential Graphs", *Axiomathes* 15, 513-540.

2005a. "IF Logic and Games of Incomplete Information", in J. van Benthem et al. (eds), *The Age of Alternative Logics: Assessing the Philosophy of Logic and Mathematics Today*, Dordrecht: Springer, 243-259.

2004b. "Grice in the Wake of Peirce", *Pragmatics & Cognition* 12, 295-315.

2004a. "Semantic Games in Logic and Epistemology", in S. Rahman, D. Gabbay, J. P. Van Bendegem, J. Symons (eds), *Logic, Epistemology and the Unity of Science*, Dordrecht: Kluwer, 57-103.

2003. "Peirce's Game-theoretic Ideas in Logic", *Semiotica* 144, 33-47.

2001c. "Propositional Logic of Imperfect Information: Foundations and Applications", *Notre Dame Journal of Formal Logic* 42, 193-210.

2001b. "Varieties of IFing", in G. Sandu and M. Pauly (eds), *Proceedings of the ESSLLI'01 Workshop on Logic and Games*, Helsinki, August.

2001a. "Intentional Identity Revisited", *Nordic Journal of Philosophical Logic* 6, 144-188.

2000. "Logic and Coherence in the Light of Competitive Games", *Logique et Analyse* 43, 371-391.

17

Graham Priest

Distinguished Professor, the Graduate Center of the City University of New York Boyce Gibson Professor Emeritus, the University of Melbourne

---

1. Why were you initially drawn to the philosophy of logic?

Well, in the first place, I was drawn to logic. I was trained as a mathematician. My doctorate was in mathematical logic, and so I engaged with some of the great results of logic of the first half of the 20th Century, such as Gödel's incompleteness theorems. This made me reflect on some of the philosophical puzzles and paradoxes in the neighbourhood of these. In this way, I was drawn into philosophical logic. I wanted to see what was going on under the mathematics.

Mathematical constructions and results are often beautiful and important in their own right, but professionally I suppose that I have rarely been interested in them for their own sake. Mathematics is important for me when it engages with philosophical issues. (Thus, I gave up reading the *Journal of Symbolic Logic* a long time ago. I have no problem with those who have interests in such areas; but any connection which the papers in this journal have with philosophical issues largely disappeared a long time ago.) In particular, "classical logic", that is the logical theory invented by Frege and Russell and co-travelers, is a superb logical tool. So much better than any logical theory that came before. But every logician is aware that it faces philosophical problems – though some might think that they can be made to disappear with appropriate manoeuvering. It seems to me that, in many cases, this is not so. It must be possible to do better. Much of my work in logic has been directed to seeing how.

2. What are your main contributions to the philosophy of logic?

For the part of my work in question, I find it hard to disentangle the philosophy of logic from logic itself, the philosophy of language, and metaphysics. I suppose that most people who know of my work will associate it with dialetheism, that is, the view that some contradictions are true. I think that in the first place, this is a view in metaphysics.

Thus, Aristotle defends the Principle of Non-Contradiction in his *Metaphysics*, not his *Analytics*. Since Leibniz, however, the Principle has been taken to be a part of logic. I have advocated dialetheism on many grounds, concerning the paradoxes of self-reference, motion, inconsistencies in law, contradictions that arise at the boundaries of what can be said/thought – and most recently in connection with aspects of Buddhist philosophy.

Of course, since dialetheism is such a contentious view, much of what has to be done in advocating it is defending it against objections. Of these, the first that comes to mind to most philosophers nowadays is the thought that everything follows from a contradiction (Explosion); so dialetheism lapses into trivialism. This is, in fact, a very superficial objection, since any dialetheist who is not a trivialist will take Explosion not to be truth-preserving and so not to be valid. What is not superficial, however, is framing a robust account of validity which shows how and why Explosion fails. Logics which invalidate Explosion are called paraconsistent logics. They were invented by various people before I came on the scene (Jaśkowski, Da Costa, Anderson and Belnap), but I formulated one in ignorance of their work (the so called Logic of Paradox), and much of my work since has been connected with developing various paraconsistent logics, especially (though not exclusively) in connection with relevant logics. I have also been engaged in looking at the application of such logics to truth, sets, arithmetic, and exploring the interesting philosophical and mathematical possibilities that arise there.

Another objection that is often made against dialetheism is that one can suppose that some things of the form $A\&\sim A$ are true only by changing the meaning of negation. This is another superficial objection because, as anyone who knows much about the history of logic will know, there have been many different theories about how negation works, and what properties it has. To assert, baldly, that the account given in "classical" logic is correct is simply to beg the question. However, it is beholden on a dialetheist to give an account of the meaning of negation. This, though, is relatively easy, given that an appropriate semantics of paraconsistent logic is under control.

Of course, there are many other possible objections to both paraconsistency and dialetheism (distinct from objections to the application of it to any *particular* area). These concern denial, rationality, belief revision, and a variety of other notions. The objections engage with issues in technical logic, the philosophy of language, epistemology, and other areas. I have discussed all these things, though this is not, I think, the place to go into matters.

When I started to advocate dialetheism, most people refused to take it seriously, and were therefore content with very superficial objections. Thinking philosophers now know that if it is to be refuted, a much more sophisticated discussion will be required. This is now starting to happen, and where it will lead, time will tell. However, as a result of this discussion, even if dialetheism were to turn out to be incorrect, we will learn much about logic, truth, negation, rationality, and many other notions in the process. Indeed, how could it be otherwise? Aristotle effectively manage to close down debate about the Principle of Non-Contradiction in Western philosophy. You cannot make philosophical progress in some area by stopping thinking about it.

3. What is the proper role of philosophy of logic in relation to other disciplines, and to other branches of philosophy?

For most Xs there is a philosophy of X: philosophy of biology, philosophy of mathematics, philosophy of history, philosophy of mind, philosophy of language, philosophy of art. Any topic of sufficient generality will throw up numerous philosophical questions. And the philosophy of X is the domain in which these things are pursued. So it is with logic. Logic is the study of what follows from what, and why. The 'why' is already a big philosophical question, and answering it forces one to engage with questions about truth, meaning, probability, rationality, as well as many more local questions, such as the nature of truth-bearers, logical constants, and so on. Digesting such questions is necessary to establish a theory of logic as well-grounded and philosophically defensible.

To the extent that logic is relevant to other disciplines, then, so is the philosophy of logic. Of course, all disciplines argue. So the correct canons of argumentation are going to be relevant to all disciplines. However, logic, and so its philosophy, has a particularly intimate connection to several disciplines: mathematics, computer science, and linguistics, in particular. Logic is informed by the application of mathematical tools, and, in reverse, throws up new structures for mathematics to analyse. (To give just one example: mathematical structures based on non-classical logics.) Logic provided the foundations of computation theory, and in reverse, the development of AI and computer-reasoning have provided fertile ground for new developments in logic. (To give just one example: non-monotonic logics.) Logic has provided the basis for various theories of linguistics, and in reverse, linguistics has provided impetus for the study of novel parts of logic. (To give just one example: various sub-structural logics.)

The relation of logic and its philosophy to philosophy in general is a particularly close one. Of course, philosophers argue, and so logic is relevant there. As I have also indicated, issues in philosophical logic relate to questions in the philosophy of language, epistemology, and metaphysics. However, I think that the connection between logic and philosophy goes even deeper than this. Time and again in the history of philosophy, we have seen logic deployed as the ground of metaphysics: it is the tectonic of Kant's *Critique of Pure Reason*; Hegel has a pan-logical metaphysics; in the *Tractatus*; Wittgenstein reads off his account of reality from Frege-Russell logic; Dumment shapes his philosophy of language on the verificationism of intuitionist logic; Kripke reads off his philosophy of language and metaphysics from his semantics for modal logic. There is something wildly over-optimistic about all these projects. However, they illustrate an important point. There is a sense in which logic provides the framework, the ground rules, for any metaphysical project. This does not mean that it determines an answer to the project, but it does put boundaries on how one can proceed. To give one simple, but obvious, further example: can we develop metaphysical theories which allow objects to behave in a contradictory fashion? The frame of an explosive logic says 'no'; the frame of a paraconsistent logic says 'yes'. In this way, then, logic is relevant to metaphysics. And behind most systems of ethics (or of values more generally), there is a usually a metaphysics. So the relevance of logic to all philosophy is there. Maybe at a distance, but it is there.

4. What have been the most significant advances in the philosophy of logic?

Wow! In two and a half thousand years of logic, East and West? There is no way that I can answer that question here, so let me just stick to the last 140 years in the West – roughly from the rise of modern logic. I will break this up (notionally) into three periods. Again it is difficult to divorce advances in the philosophy of logic from advances in logic itself, the philosophy of language, and so on.

I suppose that the most significant advances in the philosophy of logic in the first generation of logicians in question were in establishing the autonomy of logic from psychology, the analysis of the nature of quantifiers, and clearly framing the distinction between systems of proof and semantics (and so posing the question of how these things should be related).

These advances made possible the great results in metamathematics of the next generation of logicians, most notably in this context, the standard incompleteness results: Gödel's theorems, the unaxiomatizability of second order logic, the Skolem-Loewenheim theorem. There was enough material here to keep philosophers busy with questions about the philosophical implications of these results for a long time. In many ways, these debates are still going on. The amount of consensus on all of these matters – at least since the dismantling of the influence of Quine – is distinctly limited. I think that most of us would agree that these results are profound. Wherein lies their profundity is, however, still contested.

A word should be said, in this context, about the philosophy of mathematics. Many of the logical and philosophical advances in the period in question were driven by attempts to develop a viable theory of the foundations of mathematics: logicism, intuitionism, formalism. By the end of this period, all of them were generally agreed to have failed. This is philosophical progress, which may well arise due to the fact that we know what does *not* work.

The third generation of advances in logic really belongs to the development of non-classical logics. Some of these had been developed from early in the second period. But as logicians became more aware of the various problems and limitations of Frege-Russell logic, this was the time when non-classical logics blossomed, bringing with them a whole new bunch of philosophical problems. We see (amongst other things):

- The rise of modal logics and their semantics. Which modal logic correspond to which notion of necessity; and what is one to make metaphysically of world-semantics? I don't think there is as yet much consensus on these questions. The waters have been made even murkier in last 30 years by the rise of theories of impossible worlds.

- Debates about metaphysical realism and anti-realism, based on the supposed forced choice between "classical logic" and "intuitionist logic". That debate is now pretty dead, but it has morphed into the more general one of whether one should prefer a truth-conditional account of meaning, or a proof-theoretic account of meaning, which debate is still going strong.

- Novel theories of conditionality. The material conditional of Frege-Russell logic was never really a very good candidate for an ordinary conditional, though logicians were pretty happy with it for a while. However, its limitations eventually became clear. Hence we saw the invention of conditional logics, relevant logic, and the ensuing debates about whether these are any better. As far as I can tell, there is now little consensus about the conditional.

- New work on the logical paradoxes. Much of logic in the 20th Century was driven by the logical paradoxes. Theories thereof in the first half of the century tended to be based on Frege-Russell logic. In the period in question, we have witnessed the development of logics based, notably, on truth value gaps and/or gluts, and seen how they may or may not be applied to the paradoxes. These have added a whole new dimension to the debate, and, as far as I can tell, there is absolutely no consensus over this.

- Debates on vagueness. Though sorites paradoxes have been known since Eubulides, it is notable that such paradoxes were not debated in Medieval logic, nor in the modern period until the last 40 years. Since then, we have seen much discussion of the nature of vagueness. Much of this has been in connection with non-classical logics, such as supervaluation logic, fuzzy logic, and paraconsistent logic.

- The establishment of the field of non-deductive logics. After a brief period of trying to develop these on the basis of probability theory, work in the last 40 years has focused on the development of non-monotonic logics, based on semantics with a priority ordering.

- Debates about logical pluralism. Traditionally, logicians have assumed that there is one correct logic (at least, one correct deductive logic). The proliferation of different logical theories has caused some to wonder whether this is correct. Maybe some of these logical theories are right for some things; and some for others. That is a fairly recent debate, and we are still just feeling our way around it.

5. What are the most important open problems in philosophy of logic, and what are the prospects for progress?

I don't think it profitable to concentrate on particular questions. There are just too many, and several of them are inter-connected. As I indicated in my answer to the last question, there are many whole areas of philosophical logic where there is no consensus. Any developments in these areas which helped produced such a consensus (even temporarily) would be welcome.

But in what does progress in philosophy consist? Not in reaching consensus (much less in finding definitive answers). It consists in deepening our understanding. We come to understand new questions to ask; to understand how to ask old questions better; to understand new answers that are possible; to understand why old answers don't work properly, and maybe how to improve them. I am not foolhardy enough to make predictions about what is going to happen in any of the areas I mentioned, or about what new areas of inquiry might emerge. But we now have at our disposal a wider range of mathematical tools than ever before for applying to matters in logic; and people are feeling freer than ever to apply these tools. Perhaps this will just lead to a proliferation of theories, and even less consensus, but it is hard to see how this could not but deepen our understanding of many issues.

18
Stephen Read

Professor Emeritus of History and Philosophy of Logic
University of St Andrews, Scotland

---

1. Why were you initially drawn to the philosophy of logic?

I was an undergraduate in both mathematics and philosophy at the University of Keele in the 1960s. We studied a little logic in both subjects, though there was barely any attempt to connect logic with philosophy—but one course in philosophy of mathematics with Alan Treherne sparked my interest in logic enough for me to go to the Mathematics department at the University of Bristol for an intensive Masters course in mathematical logic. My doctorate at Oxford was in philosophy of language, with connections to linguistics inspired by the Chomsky boom then raging even in Oxford. Identifying a subject of 'philosophy of logic' came slowly, needing to be carved out between philosophy of language and philosophy of mathematics. In particular, logic was very much directed towards its use in mathematics or the analysis of language. It was largely a dogmatic discipline, not welcoming any philosophical interference. The first books with the title *Philosophy of Logic* are those by Quine [12] in 1970 and Putnam [11] in 1972, but what really first led me into the subject were two publishing events of 1975, of Kripke's 'Outline of a theory of truth' and of the first volume of Anderson and Belnap's *Entailment: the logic of relevance and necessity* [2]. Both were a revelation, in overturning what seemed to be dogmas.

The first dogma was that the logical paradoxes were really a dead subject. The story went that the set-theoretic paradoxes had caused a revolution in the foundations of mathematics around 1900, but were solved by the creation of axiomatic set theory by Zermelo and others; and the semantic paradoxes, distinguished from the other paradoxes by Ramsey, had been solved by Tarski by distinguishing object language from metalanguage. Kripke challenged this orthodoxy, revealing the hidden costs of implausibility of the Tarskian proposal. Kripke's ideas opened up a whole research programme in developing his own account—which was indeed programmatic, as his use of 'outline' in his title emphasized—though I was never particularly attracted by his positive proposal, which famously retained the "ghost" of Tarski's hierarchy. But it was revelatory in showing that what was the right response to the paradoxes was still a live issue.

Discovering Anderson and Belnap's work was even more instructive, not least in its immediate effect. I read their book while completing my doctoral thesis, and I was immediately weaned from philosophy of language to philosophy of logic and have never been back. What perhaps was most exciting was the thought that we could turn the tables on logic, so that instead of logic's dictating what philosophy must think (e.g., that conditionals must be truth-functional), philosophy could examine logic's presuppositions and find them wanting. At the same time, it seemed to me that Anderson and Belnap's programme had problems of its own at its base, and that things I'd learned in logic at Bristol, particularly in proof theory, could be applied to the questions Anderson and Belnap had raised about the correct account of logical consequence. Indeed, the focus on logical consequence, rather than logical truth, recorded a shift in view which, even though one can trace it back to Gentzen and Tarski, was only then becoming the norm.

2. What are your main contributions to the philosophy of logic?

At the same time as encountering these iconoclastic ideas in the philosophy of logic, I was also developing an interest in what is known as medieval logic. This is a bit of a misnomer, since it covers philosophy of logic and philosophy of language in the middle ages as much as, if not more than, logic itself. Since then, I have worked and thought and published in contemporary philosophy of logic and medieval logic about equally.

My main contributions in contemporary ideas have been in relevant logic, in the theory of paradox, and in what is now termed 'inferentialism'. The idea of relevant logic (sometimes known as 'relevance logic') is that in a valid consequence, the premises must be relevant to the conclusion. But it does not try to formalize that connection by a "relevance" filter. That would be a mistake, though one that is often made, especially by critics but even by some practitioners. Suppose that we said that an inference is valid if it is truth-preserving and the premises satisfy the "relevance" requirement of being relevant to the conclusion. Now take a truth-preserving inference which does not satisfy the relevance test, and suppose the premises are true. Then since the argument is truth-preserving, it follows that the conclusion must be true. But since the premises fail the relevance test, we are apparently not supposed to infer the conclusion—it's not a "relevantly valid" inference. That way lies madness. The main spur to my work in relevant logic was to find a better and more appropriate account of logical consequence which would reveal that really truth-preserving inferences were already relevant. This research culminated in my book *Relevant Logic* [14] (out of print but available online at http://www.st-andrews.ac.uk/~slr/Relevant_Logic.pdf), and my work since then has been guided by adherence to the belief that relevant logic gives the correct account of logical consequence.

My work on paradox is more recent, and has been inspired by my research in medieval logic. My early work in medieval logic should probably be classed as philosophy of language, trying to make sense of the specifically medieval notion of supposition. A term "supposits" for an object or range of objects if it stands for those objects in a sentence, so the role of supposition as a technical term is similar to that of reference in modern semantics, though broader, for it also covers the logical behaviour of quantified and general terms. I'd been aware, at least since the publication of George Hughes' translation of and commentary [8] on John Buridan's chapter on "insolubles" in his *Sophismata*, of the medievals' treatment of the semantic paradoxes, but felt that Buridan's analysis, though interesting as an alternative to Tarski's and Kripke's, was no more convincing than theirs. It slowly became apparent, however, that, although most scholarly work on the medieval account of paradoxes had focused on Buridan, the most seminal idea in medieval times was due to Thomas Bradwardine, writing some ten or twenty years earlier. It was with the publication of Bradwardine's *Insolubilia* in the 1320s that the whole approach to the insolubles (which include the semantic paradoxes, but also epistemic paradoxes and other puzzles) underwent a sea change. Before Bradwardine's attack on it, the dominant position was restrictionism, that the paradoxes resulted from an illicit form of self-reference. Bradwardine's devastating criticism overturned this doctrine, and his positive idea, that the paradoxes are implicitly self-contradictory in somehow asserting their own truth as well as their falsehood, is found in some form or other in almost all succeeding treatments in the middle ages. I believe that Bradwardine's idea is still viable, and have tried to develop it and apply it to a range of paradoxes in several recent publications, as well as editing and translating Bradwardine's *Insolubilia* into English [4]. One might call it the "multiple-meanings" solution, since Bradwardine claims, indeed, proves, that the paradoxes signify or mean many things besides what they overtly say.

A third area of philosophy of logic to which I've contributed is logical inferentialism, the idea that logical constants should be given a proof-theoretic rather than a model-theoretic semantics. I was introduced to the background to this idea in my studies with John Mayberry for my Masters thesis. But it lay dormant until I was faced with the challenge of giving a relevant semantics for the connectives in my analysis of the foundations of relevant logic. A recurrent complaint in the 1960s about relevance logic was that it had no semantics. The same complaint had been directed at modal logic in the 1950s, famously answered by Kripke in his possible-worlds semantics. Meyer and Routley, among others, rose to this challenge by formulating the ternary semantics for relevant logic. Just as Kripke modelled the unary modal operators with a binary relation of relative possibility, so the binary entailment operator was modelled by a ternary relation, but, ternary relations being so much less intuitive than binary ones, this analysis was not the public relations success that Kripke's work had managed for modal logic. Indeed, though model theory can yield significant technical results, it can only translate one meaning into another, or even distort it. My response was to propose giving the meaning of the relevant connectives proof-conditionally (in the final chapter of *Relevant Logic* [14]). The idea goes back ultimately to Gentzen, subsequently developed by Prawitz and Dummett. At its heart lies the proposal that meaning be given in the introduction-rules for a logical constant, which encapsulate its assertion-conditions, requiring that the elimination-rules should be justified by that meaning and so should lie in an appropriate relation of harmony with them. I've tried to spell this out in the notion of "general-elimination harmony" (a term coined by Nissim Francez and my colleague Roy Dyckhoff) — see [13].

3. What is the proper role of philosophy of logic in relation to other disciplines, and to other branches of philosophy?

Where there are two distinct activities and two different names, it's useful to use one name for the one, the other for the other, but unfortunately, there's been no consistent use of the terms 'philosophy of logic' and 'philosophical logic'. I prefer to use the former for the philosophical discipline of examining logical notions, the latter for the development of logical concepts and methods for philosophical purposes (contrasted with mathematical logic): thus philosophical logic will embrace modal logic, substructural logic, many-valued logic, mereology, and so on, whereas mathematical logic has included set theory, recursion theory and suchlike. Of course, these are not hard and fast divisions and can overlap. Philosophy of logic, in contrast, is philosophy, like philosophy of science and philosophy of art, with its distinct subject matter. Its role is to examine the concepts of logic, including consequence, quantifiers, identity, proof, model and so on. It can clearly overlap with philosophy of language, when looking at concepts like name, predicate and proposition, or with metaphysics, in examining identity, part and whole. Sharp distinctions are not important, so much as polarities of focus.

4. What have been the most significant advances in the philosophy of logic?

The most significant advance in philosophy of logic in recent decades has been the recognition that logical consequence is the predominant notion, and one worthy of close examination both as to its foundations and to its identification. Although the study of consequence goes back at least to Gentzen, and arguably much earlier (Bolzano, the medievals, indeed, to Aristotle), and alternative accounts of consequence include the proposed revisions of the intuitionists (with the identification of a specifically intuitionist notion of consequence by Heyting in 1930, though rejected by the founder of intuitionism, Brouwer, himself) and of the relevantists (stemming from Ackermann's famous paper 'Begründung einer strengen Implikation' [1] of 1956), the thought that logic is identical with logical truth lingered well into the second half of the twentieth century. The steps from logical truth to the focus on the consequence relation between two formulae, then to that between a set of formulae and a formula, and finally to its (possibly) most general form in the relation between sets of formulae (usually dubbed "multiple-conclusion" consequence) was slow and hard won. Even now the legitimacy of multiple-conclusion consequence is highly tendentious.

The attitude to logical revision has also changed dramatically since Quine posed his "deviant's dilemma" [12] in 1970: "when he tries to deny the doctrine he merely changes the subject." Possibilities have opened up in at least two dimensions. The rearguard action tried to identify logic with so-called first-order classical logic, resisting extensions to higher-order logic and to modal and other intensional logics. These are now much more readily accepted as part of the proper study of logic, with important applications both in philosophy and in other disciplines, such as computer science. They are what Susan Haack [6] called "extended logics"; in addition there are what she called deviant logics, such as intuitionist and relevant, and also now dialetheic, non-contractive and other logics which seek not to extend first-order logic, but to overturn it. Clearly, the extended logics can live happily side by side; recent proposals for a logical pluralism try to suggest that deviant logics can also co-exist, identifying different consequence relations as appropriate for different purposes, but all equally good and equally right. Whether this proposal is coherent seems to me to be a vital matter for future research.

One of the main spurs to logical revision was the implicational paradoxes, but most recent grounds for logical revision (and logical pluralism) have been the logical paradoxes. As already mentioned, research into the semantic paradoxes was re-awakened by Kripke's classic paper

[10] of 1975. The paradoxes seem to force a choice between revising our theory of truth or our logic. Tarski chose to constrain the account of truth by imposing a hierarchy of truth-predicates. Kripke at first glance appears to do neither: he retains classical logic (albeit, allowing truth to be a partial predicate) and he rejects the hierarchy, seeking a universal notion of truth. In reality, both logic and truth are compromised. The "ghost" of the hierarchy means that several semantic concepts cannot be expressed (in the object language), e.g., 'paradoxical', and even the truth predicate seems artificially constrained. Moreover, as subsequently interpreted, the classical logic of Tarski has been replaced by a three-valued logic, or a super-valuational logic whose consequence relation is non-standard.

These limitations have led to new avenues of research in recent years, and opened the possibility of further logical revision in light of the paradoxes. Even if there is less consensus about the right approach to the paradoxes, there is now a lively debate. Much of this has focused on recognition of the importance of Curry's paradox, though its full force, and its distinctness from the Liar paradox, is not always appreciated. Curry's aim in [5] was to investigate how the problems with Russell's paradox (of the set of all sets which are not members of themselves) might be replicated in a logic without negation. Take an arbitrary sentence A, and consider the set of all sets such that if they are self-membered then A. A semantic version of the paradox was identified subsequently: by diagonalization, there is a sentence C equivalent (or even identical) to 'If C is true then A'. An argument turning on the truth-equivalence, *modus ponens*, contraction (or absorption) and conditional proof establishes A, whatever A was. If A is $\bot$, the absurd sentence, then C is effectively the Liar (since not-B is equivalent to 'if B then $\bot$'); more interestingly, the argument derives triviality (arbitrary A) directly without a detour through a contradiction and the spread law (if A and not-A then B), so rejecting the spread law and possibly accepting true contradictions does not suffice to avoid triviality. Either some other logical principles must go (e.g., contraction) or (as I myself favour) the truth-equivalence must be qualified.

In medieval (philosophy of) logic, the most significant advance in recent years has been a better understanding of the medieval logical genre of obligations. There are many extant treatises on obligations, from around 1200 right through to the late Middle Ages, but scholars were very puzzled by them for many years. They are theoretical treatises giving different versions of rules for a curious form of disputation, though no records of any actual such disputations exist. In an obligational disputation, the Opponent presents a *casus* (a hypothetical situation) and a proposition, usually false in that situation; if the Respondent agrees to take part, the Opponent fires a series of further propositions at the Respondent, which the Respondent may only grant, deny or doubt, according to a strict set of rules, and depending on the initial proposition and the *casus*. What, however, was the point of such disputations, and why were the rules constructed as they were? There is still disagreement on the first question, but the second has yielded to detailed research, revealing essentially just two (or perhaps three) main types of theory, the standard theory (*responsio antiqua*—the old response) and a rival theory (*responsio nova*—the new response) formulated by reason of dissatisfaction with certain aspects of the standard theory. The old response had a dynamic nature, which led in some cases to changes of response in the course of a disputation. This seems to have unsettled some theorists, and the new response speaks to this issue, but at the cost of robbing the theory of some of its more exciting aspects. Sense has emerged out of darkness, but there is much more research needed.

5. What are the most important open problems in philosophy of logic, and what are the prospects for progress?

The times seem right for a thousand flowers to bloom, for the eclecticism of logical pluralism, if sense can be made of the idea of multiple equally good consequence relations. The main challenge is to avoid pluralism about consequence collapsing into pluralism, or even relativism, about truth. (Of course, if truth is relative, then so too is truth-preservation, and then consequence is also relative.) Beall and Restall [3] proposed to take the model-theoretic formulae for truth-preservation, that a valid consequence preserves truth in a model, and replace 'model' by 'case' or 'situation' as a way of capturing the necessity and formality of consequence. But does this relativize truth to 'truth in a case'; and if we now restrict cases to, e.g., consistent cases, or complete cases, or generalize it to inconsistent and incomplete cases, is there any justification for denying that consequence is truth-preservation in all cases, rather than just those which support classical logic, or intuitionistic logic or any but the weakest logic validated by the largest class of models?

Inferentialism has its own pluralists, often advocating a wide class of substructural logics as equally good. But Quine's challenge is ever present: if meaning is given by the logical rules, do not different rules (different logics) necessarily entail different meanings? One response is to separate the operational rules (as meaning-determining for the connectives) from the structural rules (as defining the underlying logic). There is much work to be done here, first, to produce a viable inferentialist account of logical consequence (monistic or pluralist); secondly, to show its superiority to the model-theoretic account (as well as showing the inadequacy of the latter).

Kripke's arguments in 'Naming and necessity' [9] produced a revolution across philosophy in showing that the concepts of analyticity, necessity and apriority were not obviously, indeed, perhaps not in fact, co-extensive. (His observation was not unprecedented—see, e.g., Sloman's paper [15] in *Analysis* 1967—but it was his arguments that were effective). One upshot has been to recognise sentences such as 'I am here now' as logically true, and context-sensitive expressions as requiring logical treatment. But a viable context-sensitive account of logical consequence is still wanting. The creation of two-dimensional semantics, separating the context of assessment from the context of utterance, was crucial, but the proper articulation has yet to be accepted into the mainstream.

There has been much technical work on truth theories inspired by Kripke's suggestions for dealing with paradox. The verdict is still open, however, on whether logic should be revised in light of the paradoxes, indeed, whether logic should be revised for other than strictly logical reasons, or whether the revisions should be in the account of truth, leaving logic intact. Are there really true contradictions, can or should they be saved from triviality, or do the paradoxes turn on insensitivity to contextual factors or blindness to the multiple meanings of expressions? There are many many questions to pursue. The only doubt is whether it will be possible to recognise the correct conclusions among the plethora of proposals.

Bibliography

[1] W. Ackermann, 'Begründung einer strengen Implikation' *Journal of Symbolic Logic* 21 (1956), 113-28.

[2] A. Anderson and N. Belnap, *Entailment: the logic of relevance and necessity*, vol.I, Princeton UP 1975.

[3] J.C. Beall and G. Restall, *Logical Pluralism*, Oxford UP 2006.

[4] Thomas Bradwardine, *Insolubilia*, ed. and tr. S. Read, Peeters 2010.

[5] H.B. Curry, 'The inconsistency of certain formal logics', *Journal of Symbolic Logic* 7 (1942), 115-17.

[6] S. Haack, *Deviant Logic*, Cambridge 1974.

[7] A. Heyting, 'Die formalen Regeln der intuitionistischen Logik', *Sitzungsberichte der preussischen Akademie der Wissenschaft, phys.-math. Klasse* (1930): 42-71, 158-169.

[8] G. Hughes, *John Buridan on Self-Reference*, Cambridge UP 1982.

[9] S. Kripke, 'Naming and necessity', in *The Semantics of Natural Language*, ed. D. Davidson and G. Harman, 253-355; reprinted as

Naming and Necessity, Blackwell 1980.

[10] S. Kripke, 'Outline of a theory of truth', *Journal of Philosophy* 72 (1975), 690-716.

[11] H. Putnam, *Philosophy of Logic*, Harper Collins 1972.

[12] W. Quine, *Philosophy of Logic*, Harvard UP 1970.

[13] S. Read, 'General-elimination harmony and the meaning of the logical constants', *Journal of Philosophical Logic* 39 (2010), 557-76.

[14] S. Read, *Relevant Logic,* Blackwell 1988.

[15] A. Sloman, '"Necessary", "A Priori", "Analytic"', *Analysis* 26 (1965-66), 12-16.

19

Nicholas Rescher

Distinguished University Professor of Philosophy
University of Pittsburgh, USA

---

1. Why were you drawn to the philosophy of logic?

I was drawn to logic already in my high school years, mainly by reading some logic books. In college and graduate school this interest was extended and deepened by several fine teachers, preeminently Carl G. Hempel and Alonzo Church. But the greatest influence was exerted by Leibniz, especially via Bertrand Russell's book and Louis Couturat's *La Logique de Leibniz*, a superb work of scholarship. I was convinced from the start that good work in philosophy—my paramount interest—required a solid basis in formal logic, and that logic itself needed a solid basis in philosophical reflection.

2. What are your main contributions to the philosophy of logic?

As I myself see it, my contributions fall into two groups: macro and micro. The macro-contributions consist in various efforts to provide a comprehensively coordinated systematization of sizable bodies of logical theory. The following book-length productions figure here:

- Counterfactual Logic: *Hypothetical Reasoning* (Amsterdam: North Holland, 1964).

- Imperative Logic: *The Logic of Commands* (London: Routledge, 1966).

- Many-Valued Logic: *Many-Valued Logic* (New York: McGraw Hill. 1969).

- Tense Logic: *Temporal Logic* (New York & Vienna: Springer, 1971); with Alasdair Urquhat.

- Logic of Inconsistency: *The Logic of Inconsistency* (Oxford: Blackwell, 1979); with Robert Brandom. Also: *Aporetics* (Pittsburgh: University of Pittsburgh Press, 2009).

- Inductive Logic: *Induction* (Oxford; Blackwell, 1980).

- Theory of Paradoxes: *Paradoxes* (Chicago: Open Court, 2001).
- Epistemic Logic: *Epistemic Logic* (Pittsburgh: University of Pittsburgh Press, 2004).
- Conditionalization Logic: *Conditionals* (Cambridge, MA: MIT Press, 2005).
- Theory of Apories: *Aporetics* (Pittsburgh: University of Pittsburgh Press, 2009).
- Collectivity Logic: *Towards a Theory of Collectivities* (Frankfurt: Ontos, 2011); with Patrick Grim.
- Logic of Collectivities: *Beyond Sets* (Frankfurt: Ontos, 2011); with Patrick Grim.

These books all represent ventures in systematization that conjoin a host of otherwise scattered albeit interconnected considerations into a coherently unified and perspicuous coordination of diversified material.

By contrast, there are also certain micro-contributions relating to the innovative introduction of various specific ideas and conceptual devices. These include:

- The *Rescher quantifier* and the notion of plurality assertion in quantification theory. [This quantifier is at work in such propositions as "Most *X*s are *Y*s," a step beyond the classical all/some range. It admits of both finite and infinite implementations.]
- The conception of *vagrant predicates* in predicative logic. [A vagrant predicate is one which, like "prime numbers never specifically mentioned by anyone" cannot be instantiated, and thus have "no known address." It permits the claim that $(\exists x)Fx$ despite an inability ever to instantiate it by adducing an x_0 for which Fx_0 obtains.]
- The conception of *autodescriptive systems* in many-valued logic. [In formalizing a system of logic the use of some logical machinery is unavoidable, and when a logical system is itself able to meet this need it is auto-descriptive. Not all systems have this feature — as when formalizing propositional logic requires quantifiers.]
- The conception of *plausibility* and *weakest-link reasoning* in the theory of conditionals. [The mechanisms for the systematization of counterfactual reasoning in particular requires going beyond the dichotomy of true/false to distinguish grades of plausibility for propositions. Special logico-semantical mechanisms become necessary at this point.]

- The *Rescher-Dienes implication* in nonstandard logic. [With standard logical implication $P ® Q$ obtains whenever Q is derivable from the conjunction of P with propositions that are themselves derivable from a null set of premises. With Rescher-Dienes implication this last condition is relaxed.]
- The *Rescher-Manor consequence relation* in nonmonotonic logic. [The Rescher-Manor consequence relation is a particular version of the preceding adopted for use in epistemic contexts.]
- The *Rescher-Brandom semantics* for paraconsistent logic. [In classical logic, inconsistency leads to disaster because anything whatsoever follows from contradictory premises. The Rescher-Brandom semantic allows contractions to be treated as no worse than anomalous local regularities. It permits the retention of classical logic for paraconsistent reasoning by offloading the inevitable complication into semantics.]

What is at issue throughout is a matter of particular conceptual devices that can serve as machine tools for the production of logical deliberations on larger themes.[1]

Moreover, the philosophy of logic is not disconnected from the history of logic, a field to which I have also contributed—above all with the rediscovery of the medieval Arabic theory of temporal modality.

3. What is the proper role of philosophy of logic in relation to other disciplines, and to other branches of philosophy?

The philosophy of logic is a branch of general epistemology. Its prime task is to clarify how logical theses and theories can be validated and utilized. And an important aspect of this is to investigate the issue of limits and limitations: to give an account of just what it is that logic itself can and cannot do for us.

Philosophy has to be developed as a systematically integrated whole, with every branch duly coordinated with the others. (An ethic that condemns theft remains toothless in the presence of an epistemology that precludes any prospect of identifying particular thefts.) And this is true in spades for logic with its duality of role as both a point and an instrument for philosophically cogent reasoning. For this reason, the cultivation of the philosophy of logic is a key element of the rational self-reflection that is an essential aspect of cogent philosophizing.

4. What have been the most significant advances in the philosophy of logic?

As I see it, the most significant advances in the modern philosophy of logic relate to the investigation of limit and limitations. Philosophers have been perfectly clear—at any rate since Leibniz—that purely logical theoretical deliberations cannot settle matters of empirical fact. A prime contribution of twentieth-century logic is its demonstration that this shortfall also pertains to the surely theoretical domain of pure mathematics. We now understand more clearly than ever before just what the limits of logic are, and how cogent reasoning in virtually every branch of knowledge—mathematics and indeed even logic itself—will ultimately require substantive commitments of some sort.

5. What are the most important open problems in the philosophy of logic, and what are the prospects of progress?

As I see it, the open problems relate primarily to the ongoing enlargement of the range of consideration. Over the first two-thirds of the twentieth century logic merely developed to meet the needs of the mathematician. Only more recently have the needs of the philosopher been addressed via the development of "philosophical logic." A great deal remains to be done here—especially in such areas as the logic of norms and values.

Moreover, the ongoing progress of specialization and division of labor has fragmented logic. There is great need for works of synthesis, coordination and systematization.

Logic as we traditionally have it has been predominately oriented towards the circumstances that follow necessarily from a given body of premises. And in the course of twentieth century, there arose concerns for an inductive logic aimed at conclusions secured only in the quantified manner of the calculus of probability. We do, however, need to go beyond this to a qualitative theory for drawing plausible or "reasonable" conclusions. Here some initial steps are under way but a great deal remains to be done.

Additionally, it would be good to have a synoptic and informative account of the development of philosophical logic and the philosophy of logic. For any such account would bring into closer view the need for logical instrumentalities in the resolution of philosophical issues.

To be sure, there is an acute danger that the philosophy of logic will evolve into just another isolated domain for specialists. This would indeed be unfortunate, for logic and the philosophy of logic should ideally be a generally available thought tool that is familiar to philosophers in general. Galen taught that an able physician must be competent in logic. Analogously the good philosopher should be competent in logic and—above all—in the philosophy of logic.

The great American logician C. S. Peirce insisted that logic is a categorically normative science that prescribes how we must reason. Bertrand Russell saw it as only hypothetically normative. (If you claim A you must for this very reason also claim B.) C. I. Lewis saw it a prominently hypothetical and pragmatic: "If you claim A, then you must also claim B, provided it is your objective to achieve certain ends." This sequence of views exhibits a progressively qualified and attenuated normativity. The salient challenge to the ongoing development of logic is to come to terms with this fact, recognizing that logic does not descend from the fulgurations of necessity but is a purposively geared instrumentality—a tool for achieving a variety of limited objectives in the rational systematization of information.

Endnotes

The relevant literature can be found through search-engine access to the italicized terms of the list.

C. S. Peirce: Philosophical Writings, ed. by J. Buchler (Mineola, NY: Dover Publications, 1955), p. 142.

20

Stewart Shapiro

O'Donnell Professor of Philosophy
Department of Philosophy, The Ohio State University, USA

---

1. Why were you initially drawn to the philosophy of logic?

It has been a long time; I do not think I remember. I have always been fascinated by the notion of rigorous proof, with arguments that (seem to) establish their conclusions beyond all conceivable doubt. A number of important philosophical questions suggest themselves, or at least they suggested themselves to me. Are the arguments in question rationally compelling, or does it only seem that way? If they are rationally compelling, then how? If they are not, then why do they seem to be compelling? Do the premises of a valid argument somehow guarantee the conclusion? In what sense? Do they follow of necessity? If so, how? What is the relationship between proof and language? Is it just a matter of tracking what some of the words mean? Or is there some deeper explanation for the (real or apparent) necessity of logical consequence?

I have been blessed with many wonderful teachers, all of whom contributed to my passion for philosophical logic. My interest in mathematics, and the notion of proof, was fostered in junior high school and high school, although there was no logic in the curriculum there. I had no idea that there was such a thing. My interest in logic, as such, began when I attended a National Science Foundation program for high school students, held at Ohio State University, under the direction of Professor Arnold Ross. There, I took a course in mathematical logic taught by Father Ivo Thomas of Notre Dame. I had an early encounter with Gödel's completeness and incompleteness theorems. I was hooked, at the age of 16. In 1973, I entered Case Western Reserve University, as a freshman, fascinated by logic. I did not know then that this curiosity would evolve into my main life's work, but that was only because I was not thinking that far ahead. At Case, I encountered more excellent teachers, many of whom shared a strong interest in logic, and closely related fields such as the philosophy of science and philosophy of mathematics. Howard Stein and Raymond Nelson were particularly influential. Professor Stein introduced me to the realm of axiomatic set theory.

I can trace much of my interest in the philosophy of logic specifically to the influence of my Ph.D. advisor, Professor John Corcoran, at the State University of New York at Buffalo. His passion for all matters logical is contagious, and he taught me, by example, how satisfaction comes only when one is careful, indeed meticulous. In this respect, philosophy of logic is no different than any other philosophical endeavor.

The logicians in Buffalo—from mathematics, philosophy, and computer science—had a wonderful spirit of cooperation, with many ideas passed back and forth. My list of teachers there included Harvey Friedman, John Myhill, Nicolas Goodman, Richard Vesley, Akiko Kino, John Case, John Kearns, and William Lawvere. This is not to mention an extremely talented group of graduate students in all three disciplines. The Buffalo Logic Colloquium, founded and run by Professor Corcoran, was also a regular source of inspiration. My interest in intuitionism, and to the ins and outs of classical set theory, can be traced to Professor Goodman.

Since graduate school, I have had fantastic colleagues and students, both at Ohio State University and at the Arché Research Centre at the University of St Andrews. Environments like these have continued to fuel my passion for the philosophy of logic, and stimulated whatever ideas I have managed to contribute. I am indebted to Neil Tennant, Harvey Friedman, Timothy Carlson, Crispin Wright, Stephen Read, Peter Clark, Fraser MacBride, Roy Dyckhoff, and all of my former and current students.

2. What are your main contributions to the philosophy of logic?

Probably my most influential contributions concern higher-order logic, primarily my book, *Foundations without Foundationalism: A Case for Second-Order Logic* (OUP, 1991). In retrospect, that was also the beginning of my eclectic orientation toward logic, adopting a variety of different logical systems for different purposes—a sort of pluralism that would extend considerably over the years. Many of the objections to higher-order logic (with so-called "standard semantics") begin with the observation that the consequence relation in question is inherently incomplete, in the sense that there can be no sound and complete deductive system for it. The argument has a premise that logical consequence is a (or the) canon for correct (deductive) inference. So whatever logical consequence is, it must be effective (in some sense). One cannot be expected to reason according to a non-effective consequence relation. Ought implies can. So second-order logic is not logic.

Of course, the main advocates of higher-order logic, such as Alonzo Church, George Boolos, and my advisor, John Corcoran, know about the inherent incompleteness of higher-order logic, and yet they advocate it anyway. One might conclude that they overlooked something rather obvious about logic, that its consequence relation needs to be effective. A better, and more charitable, interpretation of the situation is that members of both camps have different conceptions of what logic is. We get a pluralism about logical consequence if we further argue (or postulate) that there is something right about what each camp claims. In particular, there are aspects of logical consequence, or types of logical consequence, that are not so directly tied to canons of correct inference.

Many advocates of first-order logic assume (or argue for) a *deductivist*, or epistemic, conception of logical consequence. An argument is valid, in this sense, if there is a chain of reasoning, going via self-evidently valid steps, from premises to conclusion. This conception of consequence underlies the main technique used to advance knowledge, at least in the deductive sciences. We *deduce* theorems from certain premises. I do not deny that this is *a* legitimate conception of logical consequence. However, advocates of higher-order logic (with standard semantics) have a more *semantic* conception of consequence in mind. An argument is valid, in this sense, if its conclusion is true under every interpretation of the language in which its premises are true. This underlies the main technique used to show that a given piece of reasoning is *not* valid. To undermine an argument, one provides another argument, in the same form, that has true premises and a false conclusion. An argument is valid, in this semantic sense, if it cannot be refuted.

Gödel's completeness theorem entails that, for first-order languages, the deductive and the semantic notions of validity coincide. So, for such languages, perhaps there is not much point in arguing over which of the consequence notions is primary. Or at least the issue is less pressing when it comes to first-order languages. However, the two conceptions come apart in higher-order languages, when the semantics is standard. The so-called limitative theorems, such as Gödel's incompleteness theorem, the compactness theorem (a corollary of completeness), and the Löwenheim-Skolem theorems, show that a number of central mathematical notions, such as those of finitude, countability, minimal closure, and well-foundedness cannot be captured in first-order languages. Such notions are easily captured in second-order languages. The inherent incompleteness of the higher-order consequence relation is thus a direct result of the main *strength* of higher-order languages, namely their expressive richness. Jon Barwise put it well:

> As logicians, we do our subject a disservice by convincing others that logic is first-order and then convincing them that almost none of the concepts of modern mathematics can really be captured in first-

order logic. ("Model-theoretic logics: Background and aims", *Model-theoretic logics*, edited by J. Barwise and S. Feferman, New York, Springer-Verlag, 3-23).

One must go beyond first-order languages if we want to recapitulate the intuitive *semantic* content of various rather basic mathematical terms. Given that these basic mathematical terms *are* understood, it would seem legitimate to presuppose them in at least some semantic treatments. The expressive strength of higher-order languages is thus closely tied to the so-called limitations of first-order logic. As Alonzo Church wrote:

> ... our definition of the [standard second-order] consequences of a system of postulates ... can be seen to be not essentially different from [that] required for the ... treatment of classical mathematics ... It is true that the non-effective notion of consequence, as we have introduced it ... presupposes a certain absolute notion of ALL propositional functions of individuals. But this is presupposed also in classical mathematics, especially classical analysis ... (*Introduction to Mathematical Logic*, Princeton, Princeton University Press.)

My 1998 paper, "Logical consequence: models and modality" (*Philosophy of Mathematics Today*, ed. by Mathias Schirn, The Mind Association, Oxford, Oxford University Press, 1998, 131-156) articulates a number of different conceptions of the notion of logical consequence, some of which point toward an effective, proof-theoretic orientation and others toward a semantic, model-theoretic perspective. That attitude is reflected in most of my papers devoted to logic and the foundations of mathematics.

In recent years, my eclectic or pluralist orientation to logic has expanded. Although I fully accept the power and validity of classical logic, and classical mathematics generally, I have always been interested in alternative logics, particularly intuitionistic logic. Recently, I have become especially attracted to mathematical theories that invoke a weaker logic, and are rendered inconsistent (and trivial) if one assumes classical logic. Examples include Heyting arithmetic augmented with Church's thesis, intuitionistic analysis, smooth infinitesimal analysis, and perhaps even the inconsistent mathematics being developed in Australia and New Zealand (conducted with paraconsistent logics). Are those theories viable? Or, do they have to be dismissed, as incoherent, since they are incompatible with classical logic? Indeed, if classical logic is the One True Logic, then all such theories are contradictory.

If such non-classical theories are viable—as they seem to be—then we at least appear to have an extensive pluralism concerning logical consequence. A number of interesting questions arise, almost immediately. What does this tell us about the nature of logical consequence? If there are, in fact, a wide variety of logics, all legitimate in different contexts, then what comes of the traditional slogans that logical consequence is universally applicable, absolutely necessary, and the like? One might have thought that what makes proofs so compelling is that the logic is valid in *all* contexts. It hardly makes sense to say that a given argument, say classical reductio ad absurdum, or the law of excluded middle, is universally valid, without exception, only in some contexts, and not in others.

Does pluralism concerning logic undermine the necessity that underlies logical consequence? We are told, for example, that logical truths hold of necessity. Yet some classical logical truths (for example) are false in some of the aforementioned theories. So how can such "necessary" truths fail to hold?

The history of philosophy is full of refutations of what may be called global relativism. If everything is relative to something or other, then what of the statement of relativism itself? The present pluralism is not a global relativism, since it is limited to matters of logic. However, logic is ubiquitous. There is no perspective that does not invoke canons of reasoning. Does pluralism for logic threaten to bleed into a global pluralism, or a kind of pluralism or relativism concerning truth? What comes of the standard refutations of such views?

A closely related, and rather vexing question concerns the application of mathematics, say in science. What is the right logic to use when applying different mathematical theories, with different logics, to non-mathematical reality? Is there a different logic for that?

A closely related question concerns the appropriate logic for conducting the philosophy of logic, including the context of this very discussion. That question may not be as vexed as some of the others, however. Logicians are familiar with the distinction between an object language being studied, and a meta-language in which this study takes place. Those two need not have the same logic. How, exactly, this works, is a deep and interesting problem.

Another focus on my work is the logic for natural languages, as deployed in ordinary situations. In particular, I am concerned with the logic of vague terms. As powerful and important as classical logic is, one should remember that it was developed with focus on, and, for the most part only on, mathematics (and perhaps only on a part of mathematics). Arguably, there is no vagueness in mathematics. So one cannot draw too many conclusions concerning the logic of ordinary discourse from the overwhelming success of classical logic in understanding deductive reasoning in (most of) mathematics. My *Vagueness in Context* (OUP, 1996) gives an account of what vagueness is, and how it arises. A theory of the logic of vagueness ensues.

Other work of mine focuses on the logic required for a language with a truth-predicate, especially as that bears on philosophical positions concerning truth, such as correspondence theories and deflationism. I ponder what one is to make of the most interesting work on truth which demands logic revision. What, exactly, is it to revise one's logic? Fascinating, deep questions lie in the vicinity.

3. What is the proper role of philosophy of logic in relation to other disciplines, and to other branches of philosophy?

Of course, philosophy of logic is central to the philosophy of mathematics. These two disciplines are often run together. One would think that the philosophy of logic should be prominent in the philosophy of just about anything, or at least the philosophy of any discipline: philosophy of physics, philosophy of biology, philosophy of linguistics, philosophy of psychology, etc. One of the aims of each of these disciplines is to understand the methodology of the target practice, be it physics, biology, etc. Whatever this practice is, it surely involves deductive reasoning, and so logic is bound up with it.

Philosophy of logic is also central to the philosophy of language, and perhaps even some parts of linguistics, such as semantics. Here, one might think, the traffic should flow in both directions. Within the philosophy of logic, there is a lot of work on the meanings of logical terms. Some say that meaning is given in terms of inferential role— typically the introduction rule(s) and/or elimination rule(s) for the term in question. Others claim that meaning is given by truth-conditions or, in particular, how each such term contributes to the truth-conditions of formulas that contain it. What is the empirical significance of those claims? Are they the sort of thing that should interest a semanticist interested in natural languages? To take one example of influence in the other direction, possible-worlds semantics, developed for modal languages, has proven fruitful in at least some aspects of the semantics for natural languages.

The discipline of meta-ethics also broaches matters of philosophical logic. Some thinkers argue that ethical statements do not state facts. Yet, according to those views, there is still such a thing as ethical reasoning, and at least some of that reasoning at least looks deductive. So there is a project to make sense of what passes for logical reasoning in such projectivist or expressivist contexts.

Finally, some of the main applications of so-called philosophical logic have direct bearing on metaphysics. I have in mind studies of, for example, temporal logic and modal logic. These are often key aspects of metaphysical accounts of time and modality.

4. What have been the most significant advances in the philosophy of logic?

This is hard to say. It is often hard to determine what counts as an advance in philosophy, let alone a significant one, especially when one is engaged in the enterprise. For what it is worth, I am most impressed with the wealth of detailed work developing various logical systems, classical and non-classical, first-order and higher-order. But that is perhaps more in the way of logic itself than in the philosophy of logic. The work on temporal logic and modal logic is similarly impressive, especially the development of possible worlds semantics. As noted, that has proved useful in linguistics itself; so here we may have a lasting contribution. I am also deeply impressed by the range of work on the logic of truth and other paradox-prone semantic notions, as well as the work on the logic of vagueness. But from our present perspective, it is hard to judge the significance of this work. It takes some time to digest it and put it in perspective.

5. What are the most important open problems in philosophy of logic, and what are the prospects for progress?

This is also hard to say, from the present perspective. Most of my current interests are in the issues surrounding pluralism and relativism concerning logic. Closely related to those issues are issues about the logic for various mathematical notions, such as continuity and computability, and everyday notions, such as vagueness, and the open-texture of language generally. However, I should not elevate the problems and issues that concern me to the status of "most important open problems" in our discipline.

21

Peter Simons

Professor of Philosophy
Trinity College

---

1. Why were you initially drawn to the philosophy of logic?

My undergraduate degree was in mathematics. At the time I did not study logic, but just got on with understanding and constructing proofs. Towards the end of my study I was shocked to find that some people (constructivists, intuitionists) did not accept proof procedures such as *reductio ad absurdum* that mathematicians often use, and that they would neither affirm nor deny what seemed to me to be obvious facts, such as that there either are or are not finitely many '7's in the decimal expansion of π. Later, when I studied philosophy, I learnt how to do simple logical proofs, so I could help with undergraduate logic tutorials, and again I saw that there were several logical systems vying for the title of "correct". Among the logicians then in Manchester, there were several whose views were not standard. One was John Chidgey, who had studied relevance logic with Alan Anderson and Nuel Belnap in Pittburgh. I found many of the arguments for relevance in logic quite persuasive and so I took some trouble to investigate what they were doing, though I never aspired to make original contributions to that research. Another logician working in Manchester was Czesław Lejewski, who had studied in Warsaw before the war with Jan Łukasiewicz and Stanisław Leśniewski, and was a strong proponent of the virtues of the latter's unorthodox systems of logic. Also the presence of the late Arthur Prior and his systems of tense logic was then still felt in Manchester. In all there was an eclectic, pluralistic atmosphere around logic and an argumentative jostling among the different claims on behalf of this or that logic as being better than others. As a philosopher, I was naturally drawn to wonder about what would justify such claims, on what philosophical basis one might choose among competing systems of logic. At the same time I was reading Wittgenstein, Frege, Husserl, Quine and others on the nature of logic and its connections to other disciplines. So while I started out with an interest in phenomenology I walked smack into debates about logic, and they naturally aroused my curiosity and led me to read around.

After a while I established that my main interests lay in ontology and metaphysics, but there was the question as to what role, if any, logic might play in the determining as well as the formulation of metaphysical views. At the time (1970s) the standard view was that metaphysical questions could best be approached and solved by semantic methods. That involved taking the language of the area in question, putting it into as good a formal state as one could, ideally by axiomatization, and then applying semantic considerations and methods to see what metaphysical results came out. I cannot remember whether I was sceptical of this from the beginning, but I do recall being very impressed by several philosophers whom I either heard or read who did not take this view. The two living philosophers who impressed me in this way were Quine and Armstrong, the latter more so because he was prepared to argue that logico-linguistic considerations were secondary to ontological ones. Among those I read, I recall being very impressed by Prior's anti-platonism in *The Objects of Thought*, and by the straightforward realism of the Polish phenomenologist Roman Ingarden. It continued to be necessary therefore to articulate my view on the nature and role of logic and its relationship to ontology in order to be able to consider my views methodologically sound. I was inclined to be sympathetic to nominalism, and I knew that Lejewski, like his teachers Leśniewski and Kotarbiński, was a staunch nominalist. On the other hand most of the really great logicians, such as Bolzano, Frege, Russell, Gödel and Church, were platonists. Quine was one might say a regretful platonist, because he would have liked to get away with a nominalist ontology, but thought it would not be able to provide a sufficient basis for the mathematics needed for science. So his platonism consisted in accepting sets, and that was pretty much the standard view among logicians at the time. Tarski too, it turned out, was sympathetic to nominalism but was forced almost against his will to work with sets in order to get results. Leśniewski was much more uncompromising, but his refusal to have anything to do with sets or other abstract entities held him back from the kind of important metamathematical discoveries that Tarski produced in such abundance. So the old conflict between platonism and nominalism had a logical dimension that could not be ignored, and that weighed on my mind as well.

2. What are your main contributions to the philosophy of logic?

One respected colleague described my monograph *Parts* (1987) as a contribution to philosophical logic. The term 'philosophical logic' is not one I find easy to understand, perhaps because I never studied in Oxford. However, it may overlap in extension with the term 'philosophy of logic', so if someone thinks *Parts* belongs in part to the philosophy of logic, that would be a contribution. Personally I think mereology (which is what *Parts* is about) is neither logic nor its philosophy, but ontology, albeit pursued in part using formal methods. But then I think all good ontology concerned with fundamental concepts should be pursued in part using formal methods. So for my part I think my contributions lie elsewhere.

The notion falling within philosophy of logic in which I had a hand that is most used today is that of truth-makers. I think I was the first person to use that term in print, in a short essay published in 1982. The ideas went into a joint paper, 'Truth-Makers', that Kevin Mulligan, Barry Smith and myself published in 1984, and that contributed to the debate about how truth-bearers get to be true. The inspiration in my case was Russell's logical atomism and the more austere version of Wittgenstein, but I rejected their facts and states of affairs as truth-makers, in favour of moments, or as they are now called, tropes. A truth-maker is any entity M such that there is something that is true *because* M exists. What category it belongs to, what truths it makes true, and much more, are in general not logical or semantic questions but empirical ones. The literature on truth-makers and truth-making has grown considerably and the debates have become much more complex and sophisticated, but I still stand by the general idea and I still think our original paper gets most things right. In particular, we reject the idea, beloved of many truth-maker theorists, that every truth has a truth-maker. We later found out that the term, without the hyphen, had been in use for some time in Australian debates about Rylean dispositions, and had been coined independently by Charlie Martin. More recently, it has transpired that the idea of truth-making, though not the term obviously, was rife in medieval scholastic philosophy, and present *in nuce* in Aristotle. It is always good to find out that some of the greats got there first and that only time- and tradition-bound blindness kept one from seeing seemingly contemporary ideas in their work.

Another area on which I have worked is the grammar of logic. Ever since I came across categorial grammar in the work of Ajdukiewicz (and as present in the method of Leśniewski and the practice of Frege), it has impressed me as the key to understanding syntactic complexity. But, as Ajdukiewicz himself was forced to admit, his methods stalled when it came to explaining the binding of variables which is endemic to logic and mathematics. Being dissatisfied with all other accounts, I eventually managed to come up with a description which extended categorial grammar beyond functors and their arguments to include variable-binding operators and the contexts into which they bind. In the course of doing this I noticed that Frege had in his early writings made a clear distinction between marking and filling a place in an expression for such binding. Normally a bound variable both marks and fills its place, but Frege allowed a variable to mark a place that was filled by a constant. This is brilliant, and implicitly anticipates aspects of lambda-abstraction that were worked out much later by Church. But Frege dropped it in his later work, and it was more or less universally neglected. I think I was one of the first to clearly understand what he had been doing. Side-effects of this work were a very broad understanding of what lambda abstraction amounts to, and a recipe for eliminating variable binding from Leśniewski's language, by what turned out to be an instance of a more general theorem about the relationship between variable-binders with lambda and combinatorial languages.

One feature of truth on which nearly all Polish logicians from Twardowski onwards insisted was that truth is an absolute, non-relative feature of truth-bearers. Standard accounts of truth in model theory and logical semantics, however, taking account of modality, tense and other indexical features, relativize truth to an array of "indexes" or sets of items with respect to which a proposition is true or false. These include worlds, times, places, speakers and addressees, and more. I have defended the absoluteness of truth in such cases by abandoning the Tarski T-schema as a universal. Instead of saying that the sentence 'I love you' is true (or false), not absolutely, but relative to speaker S, addressee A and time T, I say that an utterance by S to A at T of a token of 'I love you' is true (*simpliciter*, absolutely) if and only if S loves A at T. Because the context contains variables, it is not a Tarskian T-scheme, which works at the level of sentence-types, but the utterance itself is true (or false) absolutely. The utterance is a particular, dated event (in other cases it could be a writing action, or a thought), and so congenial to nominalism.

Ever since encountering Leśniewski's systems of logic, I have admired them and the austerely nominalistic rigour behind them. One aspect of this has been a recognition that Leśniewski was right to retain the traditional idea of a name that names more than one object contra Frege and Russell. This, and the other possibility of a name that names no objects, complete the logical possibilities for terms when these stand for countables. Free logic takes in empty terms but avoids plural ones; much of post-Boolos plural logic avoids empty terms. Leśniewski has both. The work of term logic is today still largely artificially subsumed under predicate logic, but I hope in the future it will regain its freedom, and I have published several essays extolling various aspects of a flexible term logic.

Another feature of Leśniewski's logic is its ontologically non-committal understanding of quantification, a feature which irritated Quine when he discussed it with Leśniewski. Quine later wrongly and anachronistically attributed a substitutional theory of quantification to Leśniewski. Perhaps the nearest to a satisfactory interpretation is that of Arthur Prior, which refuses to "explain" quantification via anything else. On the other hand, Leśniewski stubbornly refused to "do semantics" with his logic because of his uncompromising antipathy towards set theory. So his refusal to accept ontological commitment for non-nominal quantification looks like just that: a mere refusal, without justification. I have expended a lot of (often quite fruitless) energy looking for a decent and nominalistically acceptable way to provide a non-committal semantics for Leśniewski's logic, which is in effect a simple type theory. In so doing I was looking for something that neither Leśniewski nor his student Tarski was able to provide. I now think I have a way.

Leśniewski set out to provide an antinomy-free and non-set-theoretic foundation for mathematics. But he got so bogged down in the minutiae of his logic that he never got round to it, and the systems as he developed them fall well short of providing even a logically acceptable formulation of simple arithmetic. Tarski, as we saw, "went over" to set-theoretic platonism, so the question remained as to whether a nominalistically acceptable semantic account of logic, including the idea of logical consequence defined in terms of models as in Bolzano, Tarski and Gödel, was even possible. The answer just may be "yes", provided we overcome one hurdle. Leśniewski, unlike Frege, accepted plural terms denoting several individuals. Are there, though, genuine terms which denote pluralities of pluralities? What about terms like 'The Beatles and the Stones', 'the French and the English', or 'Noah, his sons, his wife, and his sons' wives'? Having earlier rejected this possibility, like Leśniewski, I now think such terms do indeed do what they seem to do, and denote pluralities of pluralities, or as I now call them, higher-order multitudes. Given at least two individuals, there are higher-order multitudes of any order. This gives us the resources to provide models for a logic of any order, and so to "do semantics" without sets. Nominalists of a Goodmanian persuasion will reject this, but I think it is nominalistically kosher. The main formal difference from set theory turns on the non-acceptance of an empty set and the identification of any multitude with its singleton. Otherwise, the formal theory of multitudes of arbitrary order is quite like set theory, but not platonistic. This, as I see it, requires a fundamental shift in what counts as logic. If a logic which omits plural terms is less comprehensive than possible, one which omits terms for higher-order multitudes is likewise radically truncated. So I see the task going forward as one of eliciting the correct logic of multitudes, and that makes multitude theory part of logic, and not an optional mathematical add-on like set theory. This is work very much in progress.

A couple of other contributions: one is that in order to account for the logic of vague concepts I believe one must accept, like fuzzy logic, degrees of truth, but whereas fuzzy logic treats logical complexity value-theoretically, I think it should be treated supervaluationally. That results in a hybrid account of vagueness I call *supernumeration*. Finally, I would note that languages contain mass terms as well as count terms, and that the former fit especially awkwardly into the logic of the latter. I think it is better to take mass terms at face value and develop a term logic especially for them: mass logic. It then turns out that the basic notion of 'one', on which count thinking turns, is a quite complex notion definable only by quantifying terms. Like identity, unity is basically a second-order notion.

3. What is the proper role of philosophy of logic in relation to other disciplines, and to other branches of philosophy?

I will start by talking about the relation of *logic* to other disciplines. My thinking about the status of logic is Aristotelian and Leśniewskian. Logic is a tool, the tool for assessing correct inference; and it is ontologically and theoretically neutral. You should not be able to infer anything substantial from a logical truth. If you can, you have made a mistake in your choice of logic. Russell admitted that the theorem of *Principia Mathematica* according to which there is at least one individual is a "defect in logical purity". At the same time, I think that the scope of logic should be maximally wide. No form of inference, if it can be formally assessed for correctness, should be excluded. That includes not only higher-order deductive logic but also inductive logic and probability theory. The combination of these enjoins that an adequate explication of what correctness of inference consists in should not force any particular ontology on us.

As a result I see one primary task of the philosophy of logic as consisting in the search for an ontologically neutral account of what makes logic logic, and what makes a valid inference valid. As previous discussion may suggest, this is not a merely descriptive enterprise, but requires us to go beyond present paradigms. In connection with this, it is also part of the job of the philosophy of logic to explicate the terms required to give such an account, be they 'denote', 'true', 'negation', 'quantify', 'variable', or whatever. Since logic is there to normatively regulate actual inference, whether in politics or quantum mechanics, it has to be capable of showing how the actually employed styles of expression, whether everyday or regimented, fit the mathematically precise accounts of validity that modern semantics can offer. The philosophy of logic has to adjudicate whether the theoretically slim relations and functions plied in logical semantics are sufficiently akin to the facts of actual usage so as to consititute a reasonable explication. For example, an account of nominal expressions which takes them (as in so-called substitutional or truth-value semantics) as mere substitutibles in propositions rather than as expressions there to denote objects, is falling down on the job. The regimented idioms of logic cannot lose contact with the natural idioms of language or they will fail to provide an inferential standard. Nor should they be inadequate to the representation of the forms of inference found in the sciences, which include such matters as statistical inference and the manipulation of complex mathematical formulas. Many mathematical formulas found in application to empirical matters, from cosmological physics to economics, work because of the way the functions, predicates and terms in them are interpreted. This is an underdeveloped area: logic has been practised either as a tool for checking vernacular inference or as an inference engine within mathematics, but while it is fairly clear what mathematical equations, differentials, integrals and so on mean in pure mathematics, their application to matters of fact is less than transparent. Here the philosophy of logic is coterminous with the philosophies of mathematics and of science: the quantities, relations, functions etc. found in real science, manipulated through the methods of pure mathematics, and serving to deliver testable predictions, still have to be logically consistent in their manipulation, and we have to know how the terms in them connect to reality in order for them to work. This is by no means straightforward.

An area with which I am relatively less familiar but which I think deserves more attention is the application of logic within the social sciences, in particular in economics. Economists and other social scientists avail themselves heavily of mathematical and statistical methods, with their attendant modes of inference, while often rushing to idealization in a way which renders their results empirically questionable.

The issue of relevance that I encountered early in my acquaintance with logic is one which philosophers cannot, in my view, ignore. There is a somewhat schizophrenic attitude to relevance within my own views and practice, and I suspect I am not alone. The basic idea that the conclusion of a logically valid argument should be relevant in meaning to its premisses is one which intuitively commands support. That the moon is made of green cheese simply does not follow from the negation of Fermat's Last Theorem. The two things have, as we would naturally and unforcedly say, nothing to do with one another. No amount of explication via models or anything else will make this stubborn and inconvenient fact go away. On the other hand, it has proved an enterprise of more than two millennia to spell out what such relevance might consist in, and the party is not yet over. The fact that the most central relevantist propositional logic **R** is undecidable is regrettable but not decisive.

More intractable to me seems to be the way in which relevance considerations interact with predicate-logical and related forms of proposition. An extensional proposition like 'All birds are feathered' should not be represented as a quantification of a relevant conditional, since it is a mere statement of extensional inclusion. Likewise in tandem with 'Some mammals are not feathered' it appears to admit the inference 'Some mammals are not birds', yet in relevance terms this is a fallacy, despite the commonality of content carried by 'feathered': if any of the terms involved an inconsistency, it would be obvious that an irrelevant inference is involved. In practice, it seems to me, hardly any scientist, logician or mathematician actually takes the trouble to reason explicitly using a relevance logic, but simply carries on within the classical and trusts that things will come out right. It is wholly unclear to me why this should generally work so well, and the philosophy of logic should be up for giving an explanation. Simply being for or against relevance does not help here: we need an irenic solution.

As indicated in the previous section, I consider it part of the task of the philosophy of logic to assess the extent to which the use of a certain logic does or does not commit one to the existence of certain entities. My own solution is a fairly radical one (namely: never!), but that does not mean the question should not be answered in depth.

4. What have been the most significant advances in the philosophy of logic?

The discussion of truth, post-Tarski, has been one of the chief achievements of recent philosophy of logic, though there is no consensus. My own view is that too much recent truth theory has been too preoccupied with what to do about semantic antinomies, at the expense of understanding more clearly what truth is, what its (fundamental) bearers are, and how truths get to be true (so back to truth-making again). Related to that has been the fruitful discussion as to how best to give a semantical definition of logical consequence, initiated by John Etchemendy. Probably that subject is now more or less closed, which, if true, is good progress.

Another signal achievement has been the rehabilitation of higher-order predicate logic as a genuine part of logic, against the austere strictures of Skolem and Quine. The leading role in this reassessment was played by the late George Boolos, but his pleas for the acceptability of higher-order logic fell on open ears, primed not least by the recognition that, because of the Löwenheim–Skolem theorem, ordinary mathematical theories have an intended interpretation that is simply not captured adequately by first-order axiomatization. Everybody knows that the natural numbers are all finitely far from 0, but first-order Peano arithmetic does not force that to be true in all models.

I would also say that the philosophy around the arguments for relevance logic constitute a major advance, even though it is one that is incomplete. I remain relatively ignorant about the philosophy around substructural logics, so I had better not expose more of that ignorance than is decent.

Finally there is another area about which I know very little, but which appears to be of considerable practical as well as theoretical significance: the investigation of complexity and the mechanically tractable parts of logic. Obviously this is driven by automated inference, and in the computer age the significance of this can hardly be overestimated. There are indeed connections to the philosophy of relevance logic, as Belnap showed some decades ago. On the other hand, my philosophical view is that our logic should be the strongest possible: theoretical and practical limitations should not be allowed to dictate what we accept as logic. If some inferences are undecidable but valid or invalid, so be it. This sits with my view of mathematics, which is formalist: (pure) mathematics is concerned with what follows logically from collections of formal axioms. Some of these consequences will not be derivable in given proof-theoretic systems, and for others we may never know whether a given result follows from the assumptions or not.

5. What are the most important open problems in philosophy of logic, and what are the prospects for progress?

I find it in general difficult if not uncongenial to engage in prophecy, and in this case I am conscious of my own very partial knowledge of the field. So I will I am afraid have to refrain from pronouncing on *the most important* open problems and consider only the problems I would personally most like to see resolved.

One such issue is how to give an intuitively satisfying account of the validity of modal inference which does not employ the extravagant apparatus of possible worlds. It may be that the account already offered by Timothy Williamson does the job, and if so, good. If not, then let us keep looking.

Other areas in which I think we need greater clarity and resolution of open issues are several that I have mentioned above: how can we give a philosophically satisfactory account of the importance of relevance in logical inference while acknowledging that for the most part classical logic serves well enough for inferential purposes?

It would be good to have something more like philosophical closure on the question as to what is the best logical treatment of vagueness, whether of predicates or of individuals. As mentioned above, I have my own preferences, but it would be good to have more conclusive justification, whether for my theory or an alternative.

I would also like to be able to see how far we can go in the development of a logic (NB, not a mathematics!) of multitudes, which would be able to take up much of the slack left by omitting set theory from the standard tools in metalogic. There already exists a comprehensive theory of first-order multitudes, namely Leśniewski's system of ontology: the question is whether its extension to include multitudes of higher orders (including, perhaps, of transfinite order) would end up looking very like ZF set theory with a few twiddly differences, or perhaps more like Quine's interesting but tricky and under-researched NF system. At each stage of such a decision process one would be looking for philosophical arguments to justify the choices. Since I take it that the extended system *would* be logic, and not a mathematical addition to logic, that means there could be a whole lot more work for the philosophy of logic. Whether my conviction that once we have several individuals we automatically have multitudes of first and of higher orders is correct or not, this is just the beginning of the discussion.

22

Timothy Williamson

Wykeham Professor of Logic
University of Oxford

---

1. Why were you initially drawn to the philosophy of logic?

My undergraduate degree was in mathematics and philosophy, at Oxford in the years 1973-76. Naturally, doing that combination, one studied mathematical logic to quite a high level, and also thought about logic from a philosophical point of view. Michael Dummett was perhaps the most influential Oxford philosopher at that time. His imposing book *Frege: Philosophy of Logic* had just appeared, although he had not yet succeeded A.J. Ayer in the Wykeham Chair of Logic, which I now hold. Dummett made the dispute between realism and anti-realism central to philosophical discussion in Oxford, and linked classical logic with realism and intuitionistic logic with anti-realism. Classical logic accepts the law of excluded middle, according to which everything either has a property or lacks it; intuitionistic logic rejects that principle. As undergraduates, we often debated which logic was the right logic. For example, a friend and I had a long-running argument about whether intuitionistic logic solved the paradoxes of vagueness (how many hairs must you have not to be bald?): he said it did, I said it didn't.

My doctoral dissertation at Oxford was on the idea that scientific theories may approximate the truth more and more closely without ever getting there. The idea is less closely connected to the philosophy of logic than one might think, since it is perfectly consistent with classical logic: one doesn't need to revise standard logic or mathematics to compare how close different precise theories are to a given standard. The theory that the equator is exactly 39,999 kilometers long is closer to the truth than is the theory that it is exactly 39,998 kilometers long. Nevertheless, my first publication was relevant to the philosophy of logic ('Intuitionism disproved?', *Analysis*, 1982). Dummett-style anti-realists assert that all truths are knowable in principle, but there's an argument that some truths are unknowable in principle (it was first proposed, anonymously, by Alonzo Church, but the identity of its originator was revealed only long after my article appeared). I pointed out that although the argument works in classical logic, it doesn't work in intuitionistic logic, and so isn't of much direct use against Dummett-style anti-realists, because their preferred logic is intuitionistic.

Some logicians claim that the difference between classical and intuitionistic logicians is merely verbal, because they assign different meanings to the relevant logical words, such as 'not' and 'or' — they are just talking past each other. I rediscovered a result, originally proved by Karl Popper, that tells against their claim. If the dispute is verbal, then we need to disambiguate those logical words, for example by distinguishing classical negation 'not$_C$' from intuitionistic negation 'not$_I$'. The picture is that each side is right about its own words: the logic of 'not$_C$' is classical, while the logic of 'not$_I$' is intuitionistic. In particular, therefore, since double negations cancel out in classical logic, 'P is not$_C$ not$_C$ true' should entail 'P is true'; by contrast, since double negations don't normally cancel out in intuitionistic logic, 'P is not$_I$ not$_I$ true' should not entail 'P is true'. On this picture, 'not$_C$' and 'not$_I$' can co-exist peacefully in the same language, the former behaving classically, the latter intuitionistically, once they are distinguished notationally. The trouble is that the logical principles in common between classical and intuitionistic logic turn out strong enough to imply the equivalence of 'not$_C$' and 'not$_I$', so that 'not$_C$' satisfies the double negation principle only if 'not$_I$' does too. Thus peaceful co-existence is impossible. The dispute between classical and intuitionistic logic is genuine, not merely verbal. I extended that way of thinking to issues about other logical words, such as 'exist' ('Equivocation and existence', *Proceedings of the Aristotelian Society*, 1988).

Towards the end of the 1980s, I started to focus on the issue that I had debated as an undergraduate: what is the right logic for a vague language? I defended a purely classical logic as the answer, although for new reasons. Intuitionistic logic is not even the most promising alternative. Nevertheless, Hilary Putnam advocated using it to handle the paradoxes of vagueness, and I published a critique of his proposal. My criticisms had roots in those debates twenty years earlier.

As well as the specific issues that fascinate me in the philosophy of logic, I've always enjoyed the proper intellectual style of the subject: the fruitful interaction between precisely stated, rigorously proved technical results on one side and deep, elusive, non-technical philosophical concerns on the other. It has given me intense pleasure, and continues to do so.

2. What are your main contributions to the philosophy of logic?

Before answering, I'll explain what I understand the philosophy of logic to be, in order to put my own contributions to it in an appropriate setting. I take it as analogous to other branches of philosophy denoted by phrases of the form 'the philosophy of X': the philosophy of mathematics, the philosophy of physics, the philosophy of biology, the philosophy of psychology, the philosophy of economics, the philosophy of history, ... First, there is some type of inquiry: mathematics, physics, biology, psychology, economics, history, ... Then some very general questions emerge about its nature, sometimes provoked by difficulties that have arisen in pursuing the first-order inquiry itself. The questions may be proto-metaphysical: what sort of thing is this sort of inquiry about? They may also be proto-epistemological: by what sort of methodology is this sort of inquiry best pursued? When such questions are posed at a high enough level of generality, in responding to them we benefit from putting them in a broader philosophical context, to clarify the space of alternative answers and their consequences. For instance, the proto-metaphysical question may be put in the setting of theories as to what sorts of thing there are for any inquiry to be about, and the proto-epistemological question may be put in the setting of theories as to what sorts of methodology there are for any inquiry to employ. At this point, we are doing the philosophy of the sort of inquiry at issue. But we shouldn't expect the philosophy of a discipline to be somehow transcendental with respect to the discipline itself. Rather, we should expect a discipline and the philosophy of that discipline to interact strongly with each other, and even sometimes to overlap. For example, the philosophy of physics overlaps physics itself at its most theoretical.

Normally, the philosophy of a sort of inquiry concentrates mainly on the most developed inquiries of that sort we have, not on the more primitive inquiries out of which they originally grew. For example, the philosophy of physics concentrates on contemporary scientific physics, not on folk physics, which it leaves mainly to psychology. Similarly, I suggest, the philosophy of logic should concentrate on contemporary scientific logic, not on folk logic, which it should leave mainly to psychology. By 'scientific logic' I mean the sort of thing that appears in *The Journal of Symbolic Logic* and *Journal of Philosophical Logic*; it is scientific because it is governed by ideals of accuracy, systematicity, rigour, and so on, not through any special relation to experiments or observations (science isn't restricted to natural science: think of mathematics). By 'folk logic' I mean the dispositions of people formally uneducated in logic to accept some arguments and reject others. Of course, there are serious questions about the epistemology of folk logic, just as there are serious questions about the epistemology of folk physics, but we shouldn't muddle up questions about the epistemology of scientific logic with the former any more than we should muddle up questions about the epistemology of scientific physics with the latter.

Nor is scientific logic *about* folk logic, any more than scientific physics is *about* folk physics, on pain of a ruinous psychologism.

A specific difficulty for the philosophy of logic is that logic itself has an underdeveloped disciplinary identity, compared to disciplines like mathematics, physics, biology, psychology, economics, and history. In anything like its modern form it took off quite late, and departments of logic are rare, at least in English-speaking universities. It is currently divided between departments of mathematics, of computer science, and of philosophy, and in each case takes on features of the host discipline. In particular, logic driven by philosophical concerns is philosophical logic, which is in principle quite distinct from the philosophy *of* logic, although in practice the two are often not easy to distinguish. Moreover, logic consists of diverse branches which do not have very much in common with each other: think of model theory, proof theory, set theory, and recursive function theory, as well as philosophical logic. Thus it is not altogether clear what the philosophy of logic is supposed to be reflecting *on*. However, that situation presents opportunities as well as threats. We can focus the philosophy of logic in whatever way promises to be most rewarding.

My first major project in the philosophy of logic was my work on vagueness, in my book *Vagueness* (Routledge, 1994) and various associated articles. For convenience in discussing theories of vagueness, I'll use the phrase 'classical logic' broadly to cover not only principles such as the law of excluded middle but also metalinguistic principles about truth and falsity, on which the statement 'Jack is bald' is true if and only if Jack is bald, and false if and only if Jack is not bald. Most people who worked on vagueness assumed that classical logic doesn't apply to vague languages, because the dichotomy between truth and falsity breaks down in borderline cases. They thought that it had to be replaced by a non-classical alternative. The best-known candidate is fuzzy logic, although most philosophers find it rather naïve and prefer something more sophisticated, such as supervaluationist semantics. I showed that they had grossly underestimated the resources available to defend classical logic for vague languages. Given classical logic, even though 'Jack is bald' is either true or false, one would expect borderline cases in which we can't find out which, because our meanings are not perfectly transparent to us. This epistemicist view is now one of the main contenders in debates on vagueness. I showed in detail what difficulties all the alternatives get into, especially when they try to handle higher-order vagueness — borderline cases of borderline cases, and so on — which is an integral feature of ordinary vagueness. I emphasized that when one applies standard scientific criteria for theory choice, such as simplicity, elegance, strength, consistency with what is independently known, and a successful track record, classical logic stands out as the best logic even for vague languages. Classical logic doesn't need a transcendental justification, because it wins anyway by immanent standards.

After *Vagueness*, I spent some years working mainly on the epistemology that had been in its background, and wrote *Knowledge and its Limits* (Oxford University Press, 2000). However, I was also interested in the idea of absolute generality. Russell's paradox of the set of all sets that are not members of themselves is generally accepted as showing that no set can contain absolutely everything. In standard model theory, the quantifiers are always interpreted as restricted to a domain of quantification, a set, and therefore as not ranging over absolutely everything. Nevertheless, we sometimes seem to quantify over absolutely everything: for example, even when we say that no set can contain absolutely everything, it would be self-defeating to restrict our expression 'absolutely everything' to a set domain. Absolute generality also seems to be built into the ambitions of metaphysics. I explored the dialectic between absolutists and relativists about generality in a long paper ('Everything', *Philosophical Perspectives*, 2003). I defended the coherence of absolutely general quantification, while showing how close it comes to paradox, and argued that the best way of understanding it is in the setting of irreducibly higher-order logic, in which one can quantify into predicate position in a way that can't be explained in terms of the usual sort of quantification into name position.

One of the major issues discussed in my book *The Philosophy of Philosophy* (Blackwell, 2007) concerns the epistemology of logic. What entitles us to make basic deductive inferences? Unless we are entitled to make them, we can't use deduction to extend our knowledge, as we do in mathematical reasoning, for example. On one widespread philosophical view, we are entitled to make such inferences in virtue of the way in which they are built into our understanding of our logical words, such as 'if' and 'and', so that rejecting the inferences involves a kind of linguistic misunderstanding. That view, I argued, does no justice to the nature of debates over revisions of logic. The proponents of alternative logic suffer from no linguistic defect; they are just unorthodox and perhaps mistaken theoreticians. This is another manifestation of the theme that the choice between alternative logics resembles theory choice in other sciences much more closely than many philosophers of logic like to think.

My biggest contribution to the philosophy of logic may be in my book *Modal Logic as Metaphysics* (Oxford University Press, 2013). Some formulas of quantified modal logic express metaphysically very significant claims, when the quantifiers are unrestricted and the modal operators are read as expressing metaphysical possibility and metaphysical necessity. For example, one formula says that necessarily everything is necessarily something, in other words, being is non-contingent. Either that formula expresses a metaphysically significant truth or its negation does (I argue for the former). In a higher-order modal language, one can universally generalize out all the non-logical expressions in a formula and thereby extract its purely general metaphysical import. On this interpretation, modal logics can be understood as core metaphysical theories, which we must choose between on normal abductive grounds of the kind I have already explained, just as we would in other sciences. The higher-order setting is better even for evaluating theses expressible in first-order terms, because it provides a wider field in which to explore their consequences — for example, when we test their interaction with various comprehension principles for second-order modal logic, roughly speaking about which predicates define properties or relations. Although the methodology is abductive, it doesn't turn modal logic into a specifically *natural* science, because we already have enough data to be getting on with — gathering more by experiment or observation wouldn't help. In that respect, the inquiry is similar to the search in mathematics for stronger axioms of set theory, to settle the continuum hypothesis, for instance. That search also employs an abductive methodology that makes it part of general science, but not specifically of natural science: the relevant data are themselves mathematical ones. I explain how metaphysical issues arose historically in quantified modal logic, especially through the role of the Barcan formula and its converse, and discuss the virtues and limitations of possible worlds semantics, and many other issues about the application of modal and higher-order logic. The overall conception of the relation between logic and metaphysics in the book is applicable to non-modal logic too. Far from the logical positivist assumption that logic and metaphysics are mutually exclusive — because logic is good, metaphysics bad — logic (suitably interpreted) supplies the structural core of metaphysics. Logic isn't a neutral umpire between substantive scientific or metaphysical theories: it is a substantive theory itself. In some respects I'm returning to Frege's conception of logic.

3. What is the proper role of philosophy of logic in relation to other disciplines, and to other branches of philosophy?

I've already sketched my view of the relationship between the philosophy of logic and the discipline of logic. Given the intimate connections between logic and mathematics (for example, through set theory), that also yields connections between the philosophy of logic and mathematics. As I've also explained, the philosophy of logic includes the metaphysics and epistemology of logic, which are best pursued in the light of metaphysics and epistemology more generally. The epistemology of logic provides a salutary reminder to everyone concerned with human knowledge of how much we can learn by thinking rather than observation. Conversely, psychology constrains the epistemology of logic, since the latter should not rely on implausible assumptions about the reasoning processes of normal humans. There is also close interaction between the philosophy of logic and the semantics of natural languages as a branch of linguistics. To draw predictions from semantic criteria of validity for arguments in natural language, we have to combine them with semantic theories about the relevant types of sentence.

Connections with further disciplines come from the philosophy of various specialized branches of logic. For instance, the philosophy of deontic logic needs to be informed by moral philosophy and jurisprudence, since a deontic logic is the structural core of a normative theory, and for the same reason moral philosophy and jurisprudence may learn from the philosophy of logic. Similarly, the philosophy of epistemic logic can interact fruitfully with theoretical computer science and economics, which apply multi-agent epistemic logic. The philosophy of temporal logic can learn from physics, and more specifically from special relativity, concerning the status of simultaneity.

On my view, nothing in the nature of the philosophy of logic restricts us from using anything we know as evidence when relevant to the question at issue. In particular, neither logic nor the philosophy of logic is mandatorily a priori. The overall upshot is a loose holism: any discipline can in principle learn from any other, although only very rarely via a reduction of one to the other.

4. What have been the most significant advances in the philosophy of logic?

So far, the most significant advances in the philosophy of logic have been made on the back of advances in logic itself. We know vastly more about the philosophy of logic than Plato, or Aristotle, or Hume, or Kant could have done because logic has developed out of all recognition since their times. Only a tiny fraction of the relevant evidence available now was available then.

Many of the central results of metalogic themselves constitute advances in the philosophy of logic: for instance, the decidability of propositional logic, the Löwenheim-Skolem theorems, the completeness but undecidability of first-order logic, the incompleteness of second-order logic, the independence of the Axiom of Choice and of the Continuum Hypothesis, and so on. They are not simply bits of mathematics. They capture philosophically significant capabilities and limitations of different sorts of logic. If they were stated less precisely, and proved less rigorously, no one would doubt their philosophical significance: why should it vanish with an increase of precision and rigour?

Metalogic has also contributed directly to the philosophy of logic through new ideas. An obvious example is model-theoretic semantics for predicate logic, which has become the paradigm of model-theoretic semantics for a vast range of language. The introduction of a parameter for a 'possible world' opened the way for great advances in the understanding of intensional logics. On the syntactic side, the identification of a precise standard of effective computability by Church, Turing, and others made the idea of a formal system more precise and laid the foundations for the study of the complexity of logics. Gentzen's theory of natural deduction provided a new way of codifying a set of arguments, much closer than the axiomatic method to normal human reasoning. The development of non-monotonic logics for the study of non-deductive reasoning is another example, and, of course, very many more could be given.

Although my own outlook is 100% classical, I regard the multiplication of non-classical logics as progress in the philosophy of logic, because we can't have a proper reflective understanding of classical logic if we have no idea of the alternatives. More generally, even if nobody had ever actually committed themselves to a false scientific theory, we couldn't fully appreciate the status of true scientific theories without considering their rivals. If nobody had actually developed and defended non-classical logics, philosophers of logic would have been even more tempted to talk rubbish about the unthinkability of alternatives to classical logic.

The construction of interpreted formal languages of increasing expressive power has itself been a major contribution to the philosophy of logic: higher-order predicate languages, the addition of modal, temporal, epistemic, and deontic operators, plural languages, infinitary languages, and so on indefinitely. In each case the scope of philosophical reflection on logic has been significantly extended.

My examples so far have all been advances simultaneously in logic and the philosophy of logic. None of them is 'just' an advance in logic. With every one of them, if it had never been thought of, our *philosophical* understanding of logic would have been impoverished compared to its present state. Of course, an advance in logic does not simply dictate its philosophical interpretation. Reasonable people can differ in their philosophical take on it, but still it has brought the alternative pictures into clearer focus, and perhaps eliminated some or added others. At this stage, that is what we should expect.

An advance in the philosophy of logic less closely tied to a specific technical development is Tarski's 1936 account of logical consequence.

Of course, not everyone is as sympathetic to it as I am, but its simplicity, elegance, and explanatory power at the very least make it one of the salient candidates with which any rival must be compared. Work on formal theories of truth in response to the semantic paradoxes, again initiated by Tarski, casts light on the nature of truth and thereby on logic, given a connection between truth and logical consequence.

What about advances in the philosophy of logic of a more 'purely philosophical' character? Here the ground is much less firm. However, one example is the debate, going back to Quine, about the significance of real or apparent disagreement in logic. Although the overall upshot of the debate is not yet clear, we have learnt the importance of putting these issues in a general metasemantical setting, where we consider how all semantic facts are related to non-semantic facts. This is one of the places where the philosophy of logic has to learn from the philosophy of language — even though in general philosophers have greatly overestimated the intimacy of the connection between logic and language.

5. What are the most important open problems in philosophy of logic, and what are the prospects for progress?

The nature of progress in the philosophy of logic is changing, in response to changes in the nature of progress in logic itself. The greatest progress in logic has been made in mathematical logic, where the most philosophically revealing work was done fairly early on, above all in the 1930s. Since the drama of those days, mathematical logic has settled down to life as one comparatively small branch of mathematics. It might grow bigger again if it turned out to be more fruitful for other branches of mathematics than recently, but the progress involved would almost certainly still be largely mathematical rather than philosophical. Of course, people closely involved in new technical developments sometimes overestimate their philosophical significance; one must correct for the natural tendency of enthusiasm to bubble over. On the basis of the history of mathematical logic since 1970 (say), a reasonable prediction for coming decades is that for the purposes of the philosophy of logic it will merely fine-tune the overall picture we already have, however mathematically exciting the new results. Although it would be wonderful for that prediction to be falsified, I'm not on the edge of my seat. After all, major new theorems of mathematics typically lack special significance for the *philosophy* of mathematics, except on a slavishly and minutely descriptive conception of that enterprise. If someone comes up with amazing but natural new axioms for higher-order logic or set theory or some as yet unimagined alternative foundational theory, and they settle questions undecidable in current systems (such as the continuum hypothesis), I will be the first to celebrate, but that is a wild hope rather than an expectation. So we must be prepared to work in the philosophy of logic for the foreseeable future without much help from new developments in pure mathematical logic.

Recent decades have seen considerable progress in *applied* mathematical logic, by which I mean work governed by the same standards of mathematical rigour as the rest of mathematical logic but driven mainly by non-mathematical interests, which in philosophical logic include philosophical interests. Epistemic logic provides an example. The existence and utility of applied logic is a valuable constraint on the philosophy of logic, just as the existence and utility of applied mathematics is a valuable constraint on the philosophy of mathematics. However, just as one gets a distorted conception of mathematics by concentrating on applied mathematics, so one gets a distorted conception of mathematical logic, and indeed of logic in general, by concentrating on applied mathematical logic.

In my view, the most urgent task facing the philosophy of logic is to rethink from first principles the nature of logic as a scientific discipline. What is its subject matter? Many people envisage it as a study of something like idealized reasoning processes. The conception of logic as a study of idealized information processing may be a subtler version of the same idea. While idealized reasoning processes are surely worth studying in their own right, making them central to logic is a form of psychologism with which I want nothing to do. Not even the view that logic studies a priori conceptual or analytic connections is free from psychologism, for it implies a privileged role with respect to the subject matter of logic for the notional knower who attains the a priori knowledge of those connections, the thinker who possesses the connected concepts, and the member of the speech community who uses the analytically connected words.

We need a far more deeply and thoroughly realist conception of logic, on which it is no more concerned with thought, language, or information than it is with the rest of reality. When such a conception is available, even those who reject it should acknowledge that they need to engage with it if they are to make a properly reasoned case in favour of their own view. In deciding the issue, we can't simply rely on a few traditional clichés about the nature of logic. No 'pre-theoretic' stereotype of logic carries authority; we aren't going to get far by trying to give a conceptual analysis of the word 'logic'. Rather, we have to think hard about rival job descriptions for a scientific discipline of logic, which of them are most worth doing, and which are of most philosophical interest. As already mentioned, the discipline of logic as currently constituted is quite heterogeneous. That's fine. It would be ridiculous for anyone to tell all logicians what they must work on. In *Modal Logic as Metaphysics*, I specify one philosophically central task that logic is better equipped than any other discipline to carry out. At least for that purpose, what's needed is logic on the realist conception. I don't expect everyone to agree. What I'm emphasizing here is that there's a fundamental debate about the nature of logic as a discipline that we need to have, and haven't yet had because we couldn't properly see the space of possible approaches.

Of course, lots of projects in the philosophy of logic don't depend on the outcome of that foundational debate. For example, many philosophers of logic want to explore the consequences of already available conceptions of logic less realist than mine (their results may even contribute positively to the case for a radically realist conception). In the long run, however, one may hope that greater clarity about the nature of logic will help us make more progress on lots of more specific problems in the philosophy of logic.

23
Jan Woleński

Professor
University of Information, Technology and Management,
Rzeszow, Poland

---

1. Why were you initially drawn to the philosophy of logic?

I studied law (1958–1963) and philosophy (1960–1964), both at the Jagiellonian University. In my secondary school, I had very few encounters with logic. I became really interested in this subject during the first year of my legal studies. I took a quite extensive course covering propositional calculus plus the traditional logic of categorical sentences. The lectures were delivered not by a professional logician but the professor of penal law, who was a brilliant teacher. Certainly, his lectures attracted me to logic, but there was another reason, perhaps more important, for my interests in logic. I attended a very interesting introductory seminar on legal theory. Our lecturer asked us once to prepare a paper on the methodology of doctrinal study of law. I declared that I would do that. My choice was motivated by accidental and rather personal reasons. The paper was assumed to be based on a contribution published in the leading Polish journal and written by the author who happened to be a good acquaintance of my mother. She contacted us and he kindly me a copy of his PhD dissertation; the paper in question was a summary of this work. I carefully read the entire dissertation and learned many interesting things. In particular, the author referred to Alfred Tarski's idea of semantics. He quoted Tarski's famous paper, 'The Establishment of Scientific Semantics', published simultaneously in Polish and in German in 1936. I went to the library of the Department of Philosophy and asked for the *Przegląd Filozoficzny* (Philosophical Review), vol. 39, where Tarski's study was published. This paper charmed me. On that occasion I also read Tarski's 'On the Concept of Logical Consequence' (published in the same volume) with equal excitement. I began to visit this library almost every day in order to read other volumes of the *Przegląd Filozoficzny*. My readings included works by Jan Łukasiewicz, Leon Chwistek, Stanisław Leśniewski and other Polish philosophers and logicians. I also started to collect books and journals on logic. I remember that once I visited the main scientific bookshop in Kraków and bought three first volumes of the *Studia Logica*. After reading some papers I realized that my knowledge of logic was not sufficient to understand more advanced logical contributions. This strongly pushed me to study logic more extensively.

My presentation of the paper on doctrinal studies of law met favorably with the professor's expectations. He recommended that I read some books on logic and analytic philosophy. Although Poland was more liberal than other countries in the Soviet camp, we had restricted access to foreign scientific literature and very limited opportunities to study abroad or participate in international scientific collaboration. So, books and papers in Polish constituted the foundation for my student education in logic and other subjects. During my studies I became strongly interested in legal theory and the philosophy of law; my logical interests were secondary. In order to deepen my knowledge, I decided to take a second major in philosophy. Bureaucratic rules prevented me from studying philosophy starting from scratch. I had to pass examinations taken after the first year of studies, and start as a second-year student. I regretted that very much because I lost Logic One, which covered semantics and philosophy of science. Logic Two, taught in the second year, comprised an advanced course in mathematical logic. Unfortunately, our professor was a disabled person and cancelled many classes. It should, however, be said that Polish students have considerable knowledge of logic after graduating in philosophy.

Logic was not the primary subject in Kraków's philosophical environment. Roman Ingarden, one of the most important phenomenologists, was the dominant figure in this community. He considered mathematical logic to be part of mathematics and did not assign major philosophical significance to it. According to Ingarden, the genuine philosophy of logic must be placed within philosophical logic conceived according to principles of phenomenology. Due to my earlier interests in logic, my expectations (it would be hard to call them 'views' at that time), as related to philosophy of logic, were diametrically opposed to Ingarden's approach. In fact, my logico-philosophical education was strongly influenced by the Polish tradition, particularly by the Lvov-Warsaw School, which determined my route to the philosophy of logic. Logic in Poland was (and still is) understood in two ways. Logic in the broad sense (*logica sensu largo*) includes semantics (or semiotics), formal logic and methodology of science (or philosophy of science). Logic in the narrow sense (*logica sensu stricto*) is reduced to formal logic conceived as a variety of logical calculi and their metatheory (metalogic). Thus the philosophy of logic can be related either to *logica sensu largo* or to *logica sensu stricto*. Clearly, this dualism results in different collections of problems. Unfortunately, the clarity concerning the question 'What is philosophy of logic?' cannot be achieved by this distinction. In fact, there are various interconnections between particular parts of logic in the broad sense. Another problem consists in the relation between philosophical logic in the contemporary sense (the variety of logical systems constructed in order to formalize various philosophical concepts by logical means) and the philosophy of logic. Do considerations about possible worlds as the semantic machinery belong to philosophical logic or to the philosophy of logic? Inspecting contemporary textbooks, one can find different answers; some locate the problem of possible worlds in philosophical logic and some in the philosophy of logic.

I understood the philosophy of logic as related to *logica sensu stricto*; I will revisit this issue below. Polish logicians, such as Łukasiewicz, Leśniewski, and Tarski carefully distinguished logic and its philosophy and insisted that logic should not be restricted by any a priori philosophical presuppositions. For example, they sharply separated intuitionism as a philosophy of mathematics and intuitionistic logic as a formal system deserving to be studied as a regular logical system. A similar treatment was given to nominalism and Platonism as related to logic. On the other hand, Polish logicians did not deny the significance of philosophical problems of logic. Many-valued logic can be perceived as the *locus classicus* in the Polish tradition. Łukasiewicz invented three-valued logic in order to answer genuine philosophical problems related to determinism and the existence of the future. Then he generalized three-valued logic to many-valued systems with arbitrary (possibly infinite) number of logical values. At first, his hopes concerning the philosophical importance of many-valued logic were enormous. Roughly speaking, he expected that only one system, two-valued or many-valued, is satisfied in the real world. At the end of his academic carrier, Łukasiewicz became skeptical about whether this question is solvable at all. The problem, however, remains independent of Łukasiewicz's views or opinions of other logicians and can be easily identified: is logic universal or not, global or local? That is a genuine philosophical issue, which, I believe, cannot be solved via a purely formal analysis, because it combines formal and philosophical aspects. Everybody studying logic in Poland and acquainted with the Polish logical tradition learned about the philosophy of logic via many-valued logic and its philosophical consequences.

Returning to my own story, I was appointed in the Department of Legal Theory in 1963. In fact, I also received an offer from the Department of Logic. Looking back from today's perspective, I have some doubts about that choice. But counterfactual considerations about somebody's career are pointless. Anyway, I remained in close contact with the Department of Logic, particularly with Stanisław (Stan) Surma, Andrzej Wroński and Jerzy Perzanowski (he prematurely died in 2009). I participated in their weekly meetings and attended several conferences, particularly on the history of logic. In my home department one problem was hotly discussed, namely the question concerning the logic of norms. Most lawyers were inclined to think that law requires a specific logic of norms. The problem was (and still is) as follows. According to a view shared in our departments, norms are neither true nor false. On the other hand, formal logic is based on semantics with truth and falsehood as principal logical values. In particular, validity of inferences is evaluated via truth-preserving principles. Furthermore, we can point out normative arguments which look as valid in the standard sense. If norms are neither true nor false, what constitutes the semantic basis of valid inferences in which norms are involved? This puzzle is sometimes called the Jörgensen Dilemma. Another problem stems from the Ross Paradox: DoA seems to entail $Do(A$ or $B)$. However, the inference from 'Put this letter into a letter-box' to 'Put this letter into a letter-box or burn it' seems counterintuitive. Various solutions were proposed by several authors. The most radical consisted in proposing a special logic of norms, based on a specific semantics in order to solve the Jörgensen Dilemma and block the Ross Paradox. I was skeptical about such attempts. I studied writings of such authors as J. C. C. McKinsey and A. Hofstadter (incidentally, I recently learned that the latter was my remote relative), W. Dubislaw, R. Rand, K. Menger, J. Jörgensen, A. Ross, J. Kalinowski or O. Weinberger, but I did not find a satisfactory solution.

Although I was fairly enthusiastic about logical studies related to normative discourse, I decided to choose another topic for my doctoral dissertation, namely the legal philosophy of H. L. A. Hart. Several reasons motivated this choice. Firstly, I aimed to show that philosophy constitutes an integral part of legal theory, and considered Hart's ideas to be an ideal case in this respect. Secondly, I wanted to get more familiar with postwar British analytic philosophy, since I also planned to write a doctoral dissertation in philosophy devoted to the development of Wittgenstein's metaphilosophy. I planned to make a comparison between Wittgenstein's later philosophy and ordinary language philosophy in this second dissertation. (By the way, Hart himself started his career as a philosopher and belonged to the Oxford philosophical school.) I regret that, due to various personal problems, I had to abandon this project. However, my work on Hart was important for one topic belonging to the philosophy of logic. Reading writings by ordinary language philosophers (for instance, Peter Strawson), I became well acquainted with so-called informal logic. I realized that the label 'informal logic' (in the sense of Strawson and similar thinkers) refers to something very different than formal logic. Consequently, the term 'logic' is not a generic category with formal logic and informal logic as its species. As I came to see, informal logic is not equipped with clear criteria of validity. We can speak about pragmatic or rhetoric validity, and its character is significant from a practical point of view, as well as socially important. But pragmatic or rhetoric validity is still mysterious in relation to formal logical (or theoretical) categories. This lesson was very important in my own intellectual history.

I defended my dissertation (for my law degree) in 1968. After that I decided to return to the logic of norms. I also started to think about a topic for Habilitation. I choose logical problems of legal interpretation as the focus of my work. I became convinced that no specific logic of norms is required as the basis of normative inferences. According to this position, all such inferences can be tackled via deontic logic and semantics based on the concept of truth. These views were defended in my Habilitation thesis *Logical Problems of Legal Interpretation*, published (in Polish) in 1972. It contains an exposition of the view that deontic logic is sufficient for normative arguments. I extended and radicalized this position in my *Problems in Analytic Legal Philosophy* (in Polish), Uniwersytet Jagielloński, Kraków 1980. In this work I argued for two things. Firstly, since norms are not expressions, but rather decisions, i.e. facts of a kind, the logic of norm is pointless. Secondly, the standard system of deontic logic is an adequate formal basis for normative discourse. Consequently, its various extensions, which add strong permissions or conditional duties, are extralogical in principle. Such additions lead to various formal theories which go beyond logic in the strict sense.

I regarded the problem of the logic of norms as an exercise about the relation between the logical and extralogical. Because I left the legal environment in 1979 (I moved to the Institute of Social Sciences at the Technical University in Wrocław, and returned to the Jagiellonian University Institute of Philosophy in 1988), the philosophy of law was no longer the priority among my activities. In the 1980s, I worked on the monograph on the Lvov-Warsaw School (it was published in Polish in 1985 and in English in 1989) in which I included several chapters on the history of logic in Poland. This work inspired my more general interests in the philosophy of logic, going far beyond problems related to normative discourse.

2. What are your main contributions to the philosophy of logic?

If we speak about the philosophy of X, we cannot avoid the question 'What is the nature (the essence) of X?', even if the concept of nature of something is understood more or less liberally and no essentialist definition of it is postulated. Consequently, an analysis of what logic is must be regarded as the crucial task of the philosophy of logic. I do not believe that logic can be characterized *per genus proximum et differentiam specificam*. Conventionalism is the opposite approach to this issue. Certainly, we can settle our problem by adopting a convention, according to which logic consists of ...., where dots are replaced by a more or less extensive list of items subsumed under the term 'logic'. Various textbooks, handbooks, encyclopedias, companions or guides, very popular in recent times, provide examples of such fillings (the following list is not exhaustive; I remind you that I stick to *logica sensu stricto*) comprising propositional logic, first-order logic, higher-order logic, intuitionistic logic, many-valued logic, infinitary logic, IF (independent-friendly logic), Non-Fregean logic, fuzzy logic, alethic modal logic, deontic logic, quantum logic, temporal logic, etc. There are, of course, arguments for such conventions. For example, all of the logics mentioned are formal and generate patterns of inference. Of course, calculus or geometry can be constructed as formal system as well, but nobody (perhaps with the exception of traditional logicists) takes them to belong to logic.

I understand the position associated with the so-called model-theoretic approach, according to which every formalized language suitable for describing a prescribed domain (model) deserves to be called logic (this view is extensively presented in the collection *Model-Theoretic Logic*, Springer, Berlin 1985, edited by K. Barwise and S. Feferman). On the other hand, logic is expected to possess some specific properties. Now, although claims determining the scope of logic understood as *logica utens* (logic in use) are inevitably conventional and thereby, produce areas which are fuzzy to some degree, the properties of logic as a theory (*logica docens* in the traditional terminology) require something more than conventions. My initial intuition is that the philosophy of *logica docens* differs from philosophical accounts of *logica utens*. I agree that practical aspects of logic are of considerable significance, but I insist that philosophers' interests in studying the nature, whatever it is, of *logica docens* not be neglected, at least not in the philosophy of logic.

Which properties are principally relevant to logic as a theory? Pace the tradition and recent developments, I would like to mention the following general properties: consistency (eventually, paraconsistency in order to satisfy some logicians; I will, however, skip this case), axiomatizability (or another equivalent feature if logic is codified by rules), decidability, completeness (syntactic and semantic), soundness, and universality. Clearly, we expect logic to be consistent, axiomatized or reduced to a collection of rules, decidable, able to prove every formula or its negation, able to prove all truths, preserve truth (or other distinguished value or values) and be valid in all domains' models. How to analyze these attributes? I guess via metalogic (metamathematics).

Several metalogical theorems immediately mirror particular properties. For example, we can prove that first-order logic is consistent, complete, sound, axiomatizable but undecidable. This last fact has relevance to the philosophy of logic. Decidability was traditionally considered as a fundamental (Leibniz, Hilbert) property of logic and even reasonable mathematics, but Church's theorem has refuted this view. Due to this result, we can speak about degrees of logicality even within first-order logic. For instance, propositional calculus and monadic predicated calculus are decidable and Post-complete, but full first-order logic (even without identity) is not. The Lindström characterization theorem says that every logic (with Boolean negation) which is compact or complete and satisfies the Löwenheim-Skolem theorem is equivalent to first-order logic up to its expressive power. If we replace 'first-order logic' by 'logic', Lindström's result regulates the scope of logic. In particular, it excludes higher-order logic, infinitary logic or IF-logic from the domain of what is purely logical.

If we agree that the Lindström theorem characterizes *the* logic, then we accept the so-called first-order thesis (FOT) which identifies logic with the first-order system. Critics of FOT reject this identification, because it attributes only minimal expressive power to logic. I am inclined to defend FOT because it has the universality property. No metalogical results directly display this attribute, except perhaps the weak completeness theorem (A is provable from the empty class of assumptions if and only if A is true in all models). Now we have a precise explication of the claim that logic is universal in the strongest sense (provided that the empty domain is excluded). So anyone who takes the universality intuition to be important should accept FOT. In my opinion, the criticisms of FOT confuse universality as validity (truth in all models) with universality as maximal expressive power or something just short of maximal expressive power. In fact, the relation between both kinds of universality falls under the traditional principle: greater intension, lesser extension and vice versa. The weak completeness theorem is a special case of the strong one (A is provable from the set of assumptions X if and only if A is true in all models of X). This simple observation corresponds with the fact that logical systems form a subset of formal systems. (Interestingly, the Lindström result offers another general characterization of the universality property (the role of compactness is worth noting)). It does not distinguish any extralogical content (object), which is another metalogical contribution to the issue of universality. I developed the above idea in my paper 'First-Order Logic (Philosophical) Pro and Contra', reprinted in J. Woleński, *Essays on Logic and Its Applications to Philosophy*, Peter Lang, Bern 2011, 61–80. I think that this approach can be extended. For instance, since the relation of accessibility in semantics of modal logic has various properties, it means that modal systems distinguish some extralogical contents. Consequently, we should restrict modal logic to the system **K**, which does not impose any conditions on the accessibility relation. However, I would like to explicitly note that my restrictive account of what is logic, has nothing to with any depreciation of other formal or conceptual systems, in particular those intended to serve as formalizations of various philosophical notions. In fact, my collection quoted above contains papers devoted to logic going beyond the first-order paradigm. In any event, the issue of universalism (globalism) and localism in the philosophy of logic cannot be properly explored without taking metalogical results into account.

I also worked on other problems within the philosophy of logic from the metalogical point of view. I would mention three issues here. The first concerns psychologism and logic. I argue that metalogical analysis offers the most effective tools for rejecting psychologism (see 'Psychologism', in *Essays on Logic and Its Applications to Philosophy*, 31–42). The second issue pertains to the relation between syntax and semantics. The thesis that semantics has priority over syntax is not original nowadays; contemporary metamathematics provides many signs of that dependence. The priority thesis in the above sense applies to another question, namely to the relation between the formal and informal in logic. I argue (see 'What is Formal in Formal Semantics?', in *Essays on Logic and Its Applications to Philosophy*, 81–89) that formal semantics must be elaborated in an informal, though regimented, language. Thus, the formalist dream (everything can and should be formal) is just a dream, not only because of limitative theorems, but also because of the structure of human language. The priority of semantics also means that semantics cannot be exhausted via syntactic devices, even in first-order logic (the completeness theorem has no fully constructive proof). In my opinion, this fact favors realism in the philosophy of logic (see 'Logic, Semantics and Realism', in *Essays on Logic and Its Applications to Philosophy*, 51–60). Thirdly, I think that logic can be characterized as consisting of analytic and a priori sentences (see J. Woleński, 'Analytic vs. Synthetic and A priori vs. A Posteriori', in *Handbook of Epistemology*, ed. by I. Niiniluoto, M. Sintonen and J. Woleński, Kluwer, Dordrecht 2004, 781–839). As a matter of fact, the analyticity of logic is partially a defense of the analytic/synthetic distinction.

3. What is the proper role of the philosophy of logic in relation to other disciplines and to other branches of philosophy?

Since I do not believe that philosophy plays any particularly important role in any other fields, I restrict myself to the second part of the question. According to Russell, logic is the centre of philosophy. Although I do not share this view, I still consider logic to be an important source of philosophical insights. Generally speaking, the view that logic somehow displays human rationality is crucially import for metaphilosophy and favors analytic philosophy's use of logical tools. Moreover, the philosophy of logic intersects with various branches of philosophy, particularly with ontology, epistemology, philosophy of science and philosophy of language. Since logic, even in its conventional sense (see remarks above about model-theoretic logics) is a compact field, it can provide a pattern for philosophical analysis of more complex issues. However, I see no possibility of saying anything general and a priori on that topic. Determining whether philosophy of logic plays an important role in other parts of philosophy requires case studies. Several papers in my *Essays on Logic and Its Applications to Philosophy* offer such case studies. They show that there are numerous intersections between philosophy of logic and other branches of philosophy. But the principal task of these writings consisted in applying logic (not its philosophy) to philosophical analysis.

4. What have been the most significant advances in the philosophy of logic?

Doubtless, the philosophy of logic has always been dependent on the advancement of logical theories. The explosive development of logic in the 20th century gave strong impetus for philosophical considerations on logic and its nature, and demonstrated that doing philosophy of logic in isolation from formal logical tools is pointless. Although some philosophers, for instance, phenomenogists, still believe that presuppositionless philosophical analysis of logic is possible, this position is simply obsolete and unproductive. As far as that issue is concerned, the rise of various non-classical systems and generalizations of classical logic should be mentioned in this context as an essential contribution to the perennial debate about the universality of logic. Generally speaking, so-called philosophical logic has numerous and strong links with the philosophy of logic. Furthermore, as I have already noted, several basic metalogical (metamathematical) results concerning meta-properties of logical and formal systems appear to be philosophically interesting and relevant. Clarifications of various concepts traditionally used in logic, such as truth, proof, axiom, deduction, inference, model, decision procedure, complexity, derivation, logical consequence, etc. also play an important role.

5. What are the most important open problems in philosophy of logic, and what are the prospects for progress?

First of all, I expect that traditional controversies within the philosophy of logic will continue. For example, controversies about the status of objects investigated by logic and the status of logical theorems belong to this group. I do not expect very much from the formal point of view, because it seems that the most relevant metatheorems concerning logic and other formal systems have been discovered and proven. On the other hand, such predictions are risky. For instance, new characterization theorems cannot be excluded. In particular, I think that the Lindenbaum theorem (every consistent theory has maximally consistent extensions) seems to open up new possibilities. It is known, for example, that classical logic is the only maximal consistent extension of intuitionistic logic. In general, the actual role of classical logic in metamathematics should be clarified. Looking at this question from a different angle, we know that non-classical logic can be investigated by classical tools. But the problem is how far non-classical metalogic applies to classical logic remains. The same issue is even more dramatic with respect to non-classical metalogic as applied to non-classical logic. The question 'Which metatheorems about intutionistic logic are intuitionistically provable?' is a special and perhaps the most important problem of the relation between classical and non-classical metalogic. The answer seems to be relevant to the realism/anti-realism controversy in epistemology. Also, things can be suggested by cognitive science and artificial intelligence. Certainly, the study of the limits of the mechanization of inferences awaits further data. Perhaps the philosophy of applied logic will become more important than its older sister –the philosophy of *logica docens*.

Finally, the following problem seems to me philosophically interesting. It concerns the genesis of logical competence. Clearly, we have an ability to perform inferences and evaluate their correctness or incorrectness using the operation of logical consequence. And although mature logical skills characterize humans, their more primitive forms probably exist in non-human animals. Paraphrasing the metaphor introduced by Hoimar von Ditfurth, namely that the spirit did not fall from Heaven, we can say the same about logic. Yet, if logic did not fall from the Heaven (Platonic or religious), what is its natural history? In other words, how can we explain the rise of logic assuming philosophical naturalism? Furthermore, how can naturalism about logic explain the fact that logic is a priori and certain? It is possible that empiricism and psychologism in the philosophy of logic will be revived. Personally, I am of the opinion that new insights can be derived from reflections on the properties of genetic codes and information-transfers on the microbiological level. More specifically, the information space in cells looks closed by something similar to the consequence operation (one can say that the consequence operation protects information against its dispersion). On the other hand, this information space must be open in order accumulate new information. Thus, perhaps genetics and the theory of evolution can help in explaining the rise of logical competence. But this view is only philosophical speculation.

About the Editors

Tracy Alexander Lupher is an Associate Professor of philosophy at James Madison University. He is co-founder and co-Director of the James Madison University's Logic and Reasoning Institute. His work focuses on the philosophy of physics along with related issues in the philosophy of science, metaphysics, and logic.

Thomas Adajian is Associate Professor of philosophy at James Madison University. He is co-founder and co-Director of the James Madison University's Logic and Reasoning Institute. His work focuses on aesthetics and the philosophy of art along with related issues in metaphysics.

Index

A

abduction 195
aporetics 171, 172
applied logic 57, 123, 124
artificial intelligence 71, 74, 123, 124, 125, 216
autodescriptivity 171
axiomatic set theory 99, 101

B

bivalence 13, 17-19, 79, 80

C

Cantor's Theorem 56, 61
Carroll's paradox 77, 79
classical logic 2, 4, 6, 18, 19, 28, 29, 32, 39, 40, 57, 60, 67, 69, 73, 77, 79, 80, 81, 84, 88, 111, 112, 120, 131, 153, 157, 165, 166, 167, 173, 180-182, 193, 195, 196, 198, 199, 202, 215, 216
collectivities 171, 172
completeness 18, 25, 28, 34, 55, 56, 71, 88, 105, 140, 177, 179, 201, 212-214
computation 21, 22, 24, 102, 139, 155
computer science 123
consequence 2, 3, 8, 11, 21-24, 26, 28, 34, 43, 71, 74, 85, 96, 97, 137, 145, 161, 162, 164-167, 173, 178-180, 217
conservation biology 59
continuum hypothesis 13, 15, 29, 119, 200, 201, 204
Curry's paradox 4, 11, 62, 136, 161, 166

D

deduction 8, 11, 26, 79, 80, 81, 86, 90, 102, 105, 112, 125, 129, 131, 132, 199, 202, 215
definability 21
deontic logic 82, 87, 92, 109, 201, 207, 211, 212
deviant logics 22, 87, 88, 165
diagonalization techniques 105, 108
diagrams 139, 143-145
dialetheism 153-155

E

empiricism 25, 30, 32, 216
Epimenides 49, 55
epistemology 4, 31, 32, 37, 38, 40, 43, 65, 73, 77, 80, 81-85, 91, 93, 114, 134, 154, 156, 171, 173, 197, 199, 201, 215, 216
epistemology of logic 41, 77, 85, 199-201
existential graphs 96, 139
extended logics 87, 88, 165
extensionalism 105, 108

Index

F

first-order logic 4, 22, 28, 29, 79, 99, 100, 126, 140, 165, 179, 180, 201, 212-214

formalism 13, 16, 42, 74, 78, 84, 111, 113, 143, 157

Frege v, 17, 18, 23-28, 30, 32, 33, 35, 42, 46, 60, 72, 77, 81, 84, 85, 87, 88, 93, 95, 96, 99, 100, 101, 109, 144, 153, 156, 157, 185-189, 195, 200

G

game theory 139, 142, 150

Grice 139, 141, 142, 146, 149, 150

H

heterodox logic 65

heterological 49, 52, 53, 109

history of logic 23, 26, 66, 72, 130-132, 137, 154, 171, 173, 210, 211

human reasoning 99, 124, 129, 131, 133, 134, 137, 138, 146, 148, 202

I

identity 1, 10, 40, 54, 69, 71, 105, 109, 141, 143, 144, 164, 190, 195, 198, 213

IF logic 99, 100, 101, 103, 139-141

implicit/explicit distinction 21

inclosure schema 62, 137

incompleteness 25, 29, 30, 34, 37, 60, 61, 72, 100-102, 105, 109, 114, 153, 156, 177-179, 201

inductive logic 65, 71, 73, 171, 174, 190

inference 23, 26, 29, 32-34, 68, 69, 77, 79, 80, 82, 84, 85, 88, 101, 102, 106, 108, 110, 112, 116, 117, 121, 162, 178, 179, 190-193, 210, 212, 215

inferentialism 161-163

insolubles 161, 163

intensionalism 105

intuitionism 18, 23, 88, 157, 165, 177, 178, 195

K

knowability 83, 195

L

language 1-3, 5, 10, 23-25, 28, 29, 31-34, 42, 53, 54, 56, 66, 67, 70, 77-79, 82-85, 91, 94, 97, 101, 106, 107, 111-116, 121, 123, 125-127, 129, 141-148, 153-156, 161-166, 177, 179, 181-183, 186, 188, 191, 196, 200-204, 210-215

law of excluded middle 13, 18, 60, 120, 181, 195, 198

legal and scientific reasoning 87

legal theory 207-210

Leśniewski 185-189, 194, 207, 209

liar paradox 57, 60, 61, 115, 116

limitative theorems 179, 207, 214

logical consequence v, 4, 5, 8, 9, 18, 25, 28, 37, 38, 42, 84, 135, 137, 162-168, 177-181, 189, 192, 202, 203, 215, 216

logical dynamics 21

logical knowledge 41, 77-82, 85

logical non-apriorism 37, 38, 40, 45, 63

logical pluralism 3, 19, 23, 37-43, 80, 84, 158, 165, 167

logical system 29, 68, 71, 99, 101-103, 126, 172, 209

logic and probability 21, 24, 88, 190

logic of norms 174, 207, 210, 211

logic of space-time 65

M

mathematics 13-18, 23-34, 37-43, 57, 59, 65, 66, 68, 70-74, 79, 82, 84, 85, 94, 100, 102, 103, 106, 108, 113-115, 119, 121, 123, 124, 129-131, 134, 140, 146, 148, 153, 155, 157, 161, 174, 177-182, 185-187, 189, 191, 193-198, 200-204, 208, 209, 213

medieval logic 129, 130, 132, 161-163

metalogic 194, 195, 207

metaphysics 4-6, 10, 37, 38, 40, 43, 71, 84, 91, 114, 123, 134, 147, 148, 153, 156, 164, 183, 186, 199, 200, 201, 219

methods in philosophy 83, 92, 129

modal logic 57, 60, 83, 84, 87, 88, 90, 96, 97, 123, 134, 140, 143, 144, 156, 157, 164, 183, 199, 200, 212, 214

model theory 5, 22, 83, 164, 177, 188, 198, 199

Montague's Thesis 49

multitudes 185, 189, 194

N

naturalism 119, 216

nominalism 37, 42, 55, 77, 185, 186, 188, 209

non-classical logic 1, 3, 5, 6, 37, 57, 79, 153, 216

non-existent objects 105

non-reflexive logic 65, 69, 70, 73

normative/descriptive distinction 21

normativity 77, 81, 82, 85, 86, 146, 175

O

obligations 166

ontology v, 4, 37, 77, 82, 85, 147, 185, 186, 187, 190, 194, 215

P

paraconsistent logic 4, 37, 39, 65, 66-69, 153, 154, 156, 158, 173

paradox 1, 4, 8, 10, 11, 53, 55-58, 60-65, 77-79, 83, 93, 105, 108, 109, 115, 116, 135, 136, 161-163, 166,

168, 183, 199

paradoxes 129

Peirce v, 23, 24, 60, 68, 87, 92, 93, 95, 96, 97, 100, 101, 109, 139, 140-146, 149, 150, 175

phenomenology 185, 207, 208

philosophy of mathematics 13, 15, 30, 31, 37, 38, 40, 42, 57, 73, 82, 85, 94, 100, 113, 114, 119, 121, 146, 155, 157, 161, 177, 182, 197, 203, 204, 209

philosophy of science 16, 30, 37, 40, 41, 66, 82, 91, 93, 94, 114, 129, 164, 177, 208, 215, 219

platonism 13, 15, 42, 55, 186, 189, 209

pluralism 3, 13, 19, 23, 37, 38, 39, 40, 42, 43, 80, 84, 90, 147, 158, 165, 167, 177, 178, 179, 181, 183

practice-based philosophy of logic 21, 24

pragmatism 14, 44, 139, 147, 150

principle of uniform solution 61, 62, 64

probability 21, 24, 59, 60, 71, 74, 78, 83, 85, 88, 155, 158, 174, 190

probability, Classical 60, 74

probability, non-classical 59, 60

product/process duality 21, 22

proof theory 22, 126, 162, 177, 198

psychology v, 24, 81, 85, 86, 116, 119, 130, 133, 134, 156, 182, 197, 198, 201

Q

quantification 37, 40, 52, 55, 70, 79, 85, 99, 107, 144, 145, 172, 188, 189, 192, 199

quantum logic 13, 16, 17, 19, 73, 212

quasi-truth 38, 67, 69-73

R

reasoning 7, 16, 24, 25, 32, 34, 35, 37-44, 52, 59, 61, 74, 82, 85, 87, 92, 93, 96, 99, 101, 102, 105, 107, 110-116, 120, 124-138, 142-146, 148, 155, 172-174, 179-183, 199-204

reasons 29, 51, 56, 57, 77, 82, 130, 142, 145, 168, 196, 207, 210

relations 2, 3, 9, 10, 21, 23, 25, 27, 31-34, 39, 93, 100, 105, 110-116, 120, 146, 147, 164, 165, 167, 190, 191, 200

relativism 39, 80, 147, 167, 177, 181, 183

relevance logic 6, 11, 162, 164, 185, 192, 193

Rescher quantifier 172

Russell's paradox 49, 53, 56, 61, 62, 166, 199

S

second philosophy 119

semantics 5, 26, 55, 68, 70, 77, 85, 100, 101, 107-109, 112, 115, 123, 126, 140-

148, 154-158, 163, 164, 168, 173, 177-179, 182, 183, 188-191, 198, 200-202, 207-214

set theory 10, 13, 15, 18, 25, 29-31, 34, 42, 53, 56, 68, 72, 99-103, 119, 161, 164, 177, 178, 189, 194, 198, 200, 203

sorites paradox 57-62

styles of reasoning 37, 41-44

T

theory of valuations 67, 70, 71

tolerance 18, 59, 60, 80

topic-neutrality 1, 9

topological generalization of the sorites 58

topos theory 13, 15, 19

truth 1-11, 15-19, 28, 29, 33-35, 38, 49-54, 56, 67, 69-77, 83-97, 101, 106-112, 115, 119, 123, 131-136, 154-158, 161-169, 181-183, 187-192, 195, 198, 200, 203, 210-215

U

universal 1, 2, 9, 56, 107, 109, 116, 121, 147, 166, 188, 209, 213

universality 207, 212-215

V

vagueness 57-60, 79, 81, 88, 120, 158, 182, 183, 190, 193-198

validity 4-9, 15, 28-32, 49, 55, 56, 90, 116, 137, 154, 179, 180, 185, 190, 193, 201, 210-213

W

weakest-link principle 172

Y

Yablo's Infinite Liar 49, 55

www.ingramcontent.com/pod-product-compliance
Lightning Source LLC
Chambersburg PA
CBHW021839220426
43663CB00005B/323